C0-AJX-091

# 80
# WOODCRAFT
# PROJECTS

# 80 WOODCRAFT PROJECTS

by

Ronald P. Ouimet

JONATHAN DAVID PUBLISHERS, INC.
MIDDLE VILLAGE, NEW YORK 11379

**80 WOODCRAFT PROJECTS**

Copyright © 1980
by
Ronald P. Ouimet

No part of this book may be used
without the written consent of
JONATHAN DAVID PUBLISHERS, INC.
68-22 Eliot Avenue
Middle Village, NY 11379

**Library of Congress Cataloging in Publication Data**

Ouimet, Ronald P.
80 woodcraft projects.

1. Woodwork. I. Title.
TT180.092 684.1'042 80-22020
ISBN 0-8246-0260-9

Printed in the United States of America

684.1042
O93e

This book is dedicated to my outstanding parents, Anita and Roland Ouimet.

# ACKNOWLEDGMENTS

My grateful appreciation to the following people whose direct and indirect influence made this book possible:

Mark Visciano for skilled craftsmanship on a number of projects; David Mendelsohn for expert photography; Lorna Healey for the use of her child's cradle; Maurice Ouimet, Dr. J.B. Morgan, and Erick Sargent for their guidance, instruction, and inspiration.

I thank my wife and two daughters, who patiently lived with me while I worked on this book.

# Contents

# FOREWORD

This collection of major woodcraft projects is geared for the woodworker with some degree of skill and experience. And yet, it has been my purpose to keep the project construction easy to follow. The items included express fine design, including modern and American period styling. Wherever possible, time-saving construction techniques have been presented.

The projects in this book range from items that can be made using only basic tools and requiring only a few hours time to more difficult pieces involving complex cuts and joinery. Because the lathe, a specialty machine, is not always available to the home craftsman, this book contains a minimum of woodturning projects. Included, however, is a short section containing woodturning hints.

The instructions for each project reflect what I believe is the simplest way to fabricate it while insuring strength and durability. Compromise with construction techniques has been made only when it seemed it would be advantageous to the greatest number of woodworkers.

A craftsman usually develops his or her own manner of working with wood. Therefore the craftsman is encouraged to modify each project to suit his own individual taste. This will result in a feeling of greater satisfaction with the completed project.

The construction procedures given include recommendations for fasteners that might be used—screws, dowel pegs, nails, glues, etc. These are only suggestions, however, and if the craftsman believes a more detailed joint should be used, the choice is his.

In assembling this book it has been my main desire to offer the reader/craftsman a sense of pride and fulfillment in his or her woodworking. If I have achieved even a small part of that goal, I have succeeded.

Ronald P. Ouimet

October 15, 1980
Greenland, New Hampshire

# HOW LUMBER IS CLASSIFIED

Lumber is classified either as softwood or hardwood. Hardwood comes from broad-leaved deciduous trees. Some examples are oak, mahogany, chestnut, elm, birch, maple, and cherry. Hardwood is usually more expensive than softwood, but it also has a more attractive grain, making it excellent for use in cabinetmaking and furniture construction. Softwoods come from needle-bearing, evergreen trees. Typical softwoods are pine, spruce, fir, cedar, and redwood. Softwoods are used to make some furniture, although most softwood is used in house construction because of its moderate cost and easy working characteristics.

## How Lumber Is Worked

If lumber is bought just as it comes from the saw mill, the surface of the lumber is rough. Before it can be used for building furniture, it must be *surfaced* smooth by running it through a machine called a *planer*. Lumber sold in a lumber yard has usually already been smoothed (surfaced). Lumber can be bought surfaced on two sides (S2S) or four sides (S4S).

## Understanding Lumber Sizes

When purchasing 2 by 4 lumber (2 inches by 4 inches) from a lumber yard, you will notice that the actual measurements of the wood are 1⅝ inches by 3⅝ inches. The rough size board was 2 inches by 4 inches, but its dimensions were reduced when it was surfaced. The same holds true for all lumber. The actual measured size of a 1-inch board is ¾ inches.

## How Lumber Is Seasoned

A newly cut log is green and can contain up to 70 percent moisture. The logs are rough cut into boards and stacked one on top of another with space in between. They are placed in the open air to dry, one year's time for each inch of thickness of board. Lumber dried in this manner is called *air-dried lumber* (AD). When fully dry, the lumber retains approximately 12 to 18 percent moisture. A better way to season lumber is to place it in a special drying room similar to a large oven called a *kiln*. Kiln-dried lumber (KD) is better for use in furniture projects, for it has only 4 to 10 percent moisture content.

## Grading of Hardwood

The finest grade of hardwood is FAS, meaning "first and seconds." This grade is best used in fine furniture construction. Grade number 1 and grade number 2 have some defects and are lower in overall quality than FAS.

## Grading of Softwood

The select grades of softwood are grade A and grade B. A and B grades of softwood are used for some furniture projects and for the trim on the inside of a house. C and D grades of softwood have more knots and other defects, and they are also cheaper. Common softwood lumber is only used for rough construction-type work.

## Ordering Your Lumber

In ordering your lumber always specify the following:

1. The number of pieces desired.
2. The size of the stock desired.
3. The kind of wood desired.
4. The grade of lumber desired.
5. The kind of surfacing desired (rough, S2S, or S4S).
6. The kind of seasoning desired (air-dried or kiln-dried).

## Plywood

Plywood is sold in 4 by 8-foot sheets in thicknesses of ⅛ inch to ¾ inch. For light uses, such as for the back of a cabinet, the ¼ inch up to ½ inch thickness will be ample. If a project is to be made for outdoor use, use the exterior grades of plywood; the glue used to bond the layers is waterproof.

Plywood surfaces are graded A, B, C, and D. The A grade represents the smoothest and best grade while the D grade has the most defects and is the least desirable. AA (best grade both sides) is used for projects in which both surfaces of the wood will be exposed. Use the less expensive grades for project parts such as shelves, bottoms, backs, etc.

### Treatment of Plywood Edges

The end grain of plywood is layered, giving it an unattractive appearance. It is best to conceal the end grain, and there are a number of ways to achieve this. One way is to miter the corners, as a mitered joint will hide the end grain. Another method is to use wood veneer tape (refer to drawing). The veneer tape can be purchased in most wood grains. The tape is nothing more than a very thin strip of wood approximately ¾-inch wide. The tape is simply glued to the plywood

WOOD VENEER TAPE

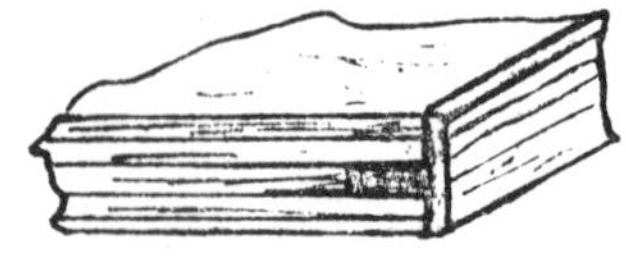

METAL OR PLASTIC EDGE

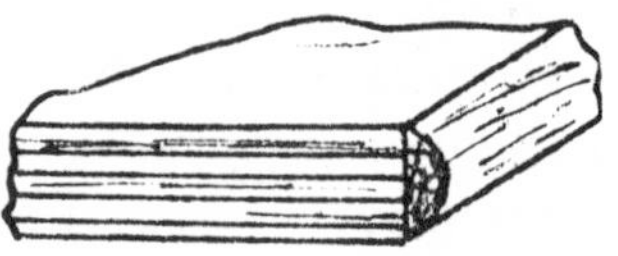

HALF ROUND WOOD MOLDING

WOOD MOLDING

edges by applying it evenly with hand pressure. Be sure to align it correctly.

Another method of concealing the plywood edge is to use matching molding. Molding will provide a decorative edge. Molding can also be purchased in metal or plastic, which is desirable for an edge that will receive abuse.

## Hardboard

Standard hardboard can be purchased in 4 by 8-foot sheets ⅛ to ¼-inch thick. Hardboard is excellent for use in general cabinetwork, such as for drawer bottoms, cabinet backs, etc. Hardboard can be purchased with holes spaced approximately 1 inch apart. *Perforated hardboard,* with the proper hooks, is excellent for storing tools and equipment. Perforated hardboard is also good for the backs of stereo cabinets. For furniture that is to be placed outdoors or will in any way be exposed to mositure use *tempered hardboard.*

Other members of the hardboard family are *chipboard, flakeboard,* and *particleboard.* These hardboards can be purchased in a ¾-inch thickness, and they have a much coarser grain.

One disadvantage of working with hardboard is its lack of holding power with fasteners. Screws and nails will stand up to more pressure when in solid wood or plywood than in hardboard.

# HARDWARE

## Screws

When purchasing screws, specify the length and diameter desired. Numbers 1 to 16 indicate the

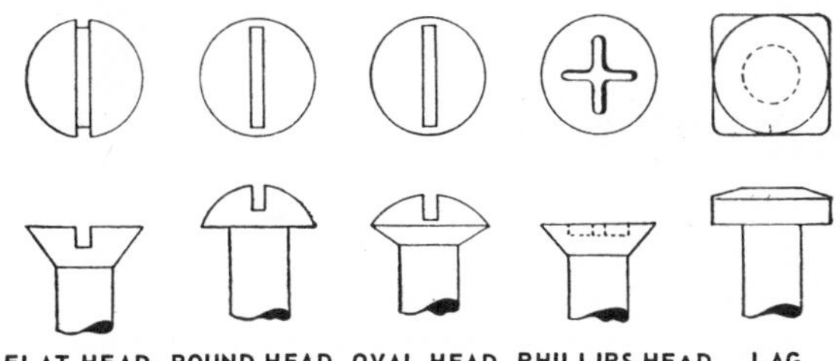


*Woodscrew head*

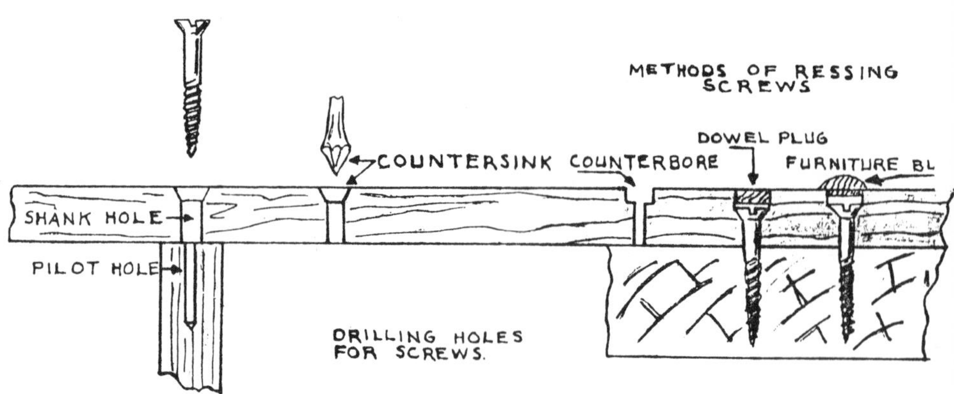


diameter. If the screw shank is a #16, it is the largest diameter; if it is #1, it is the smallest. (The drawing shows the most common woodworking screws.) When drilling into plywood or hardwood, always use a pilot hole. Use a shank or clearance hole in the first piece of

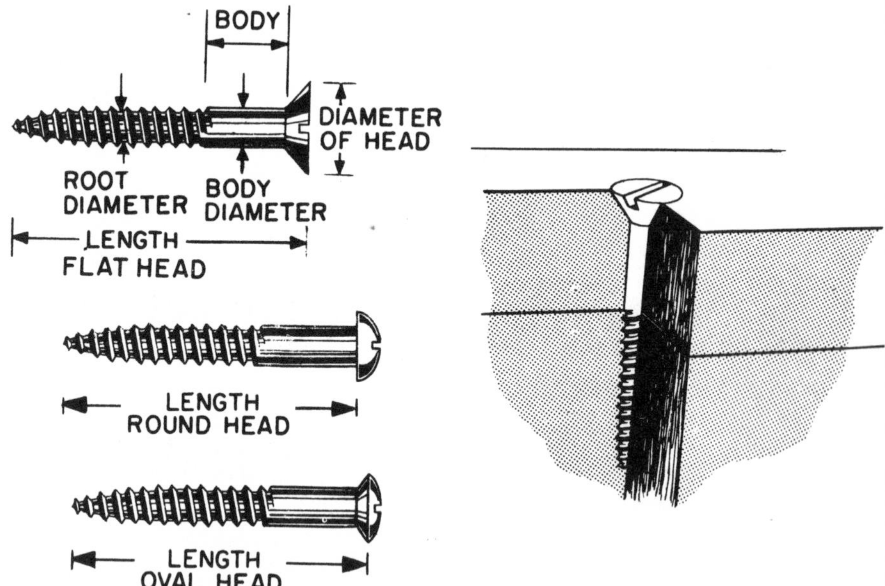


*The diagram to the left shows the nomenclature and use of woodscrews. The diagram to the right shows how a wood-screw can be used to hold two pieces of wood together.*

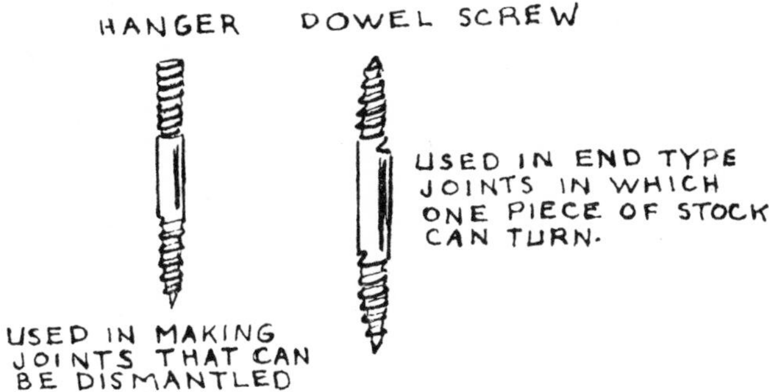


wood when screwing two pieces of stock together. The shank or clearance hole allows the two pieces to fit together tightly. Without the shank hole the first piece tends to "stay out" from the second piece, preventing a tight fit. For appearance it is always wise to set a flathead screw into a countersink or counterbored hole. If a counterbored hole is used, fill it with a furniture button or dowel plug. (Refer to drawing.)

## Nails

When purchasing nails, refer to the size as "penny" (abbreviated D).

*Common* nails have heavy, flat heads and are used

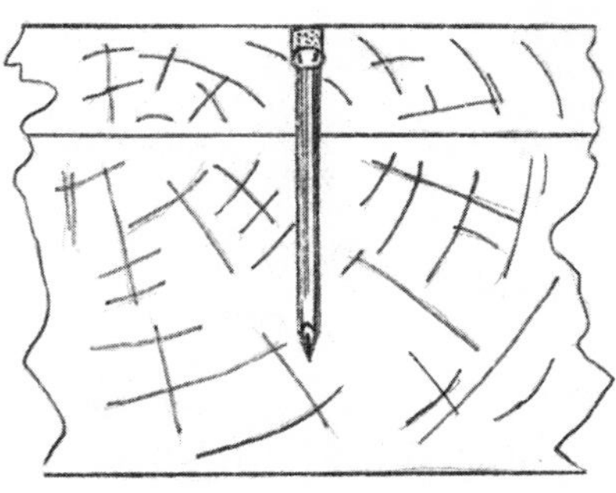

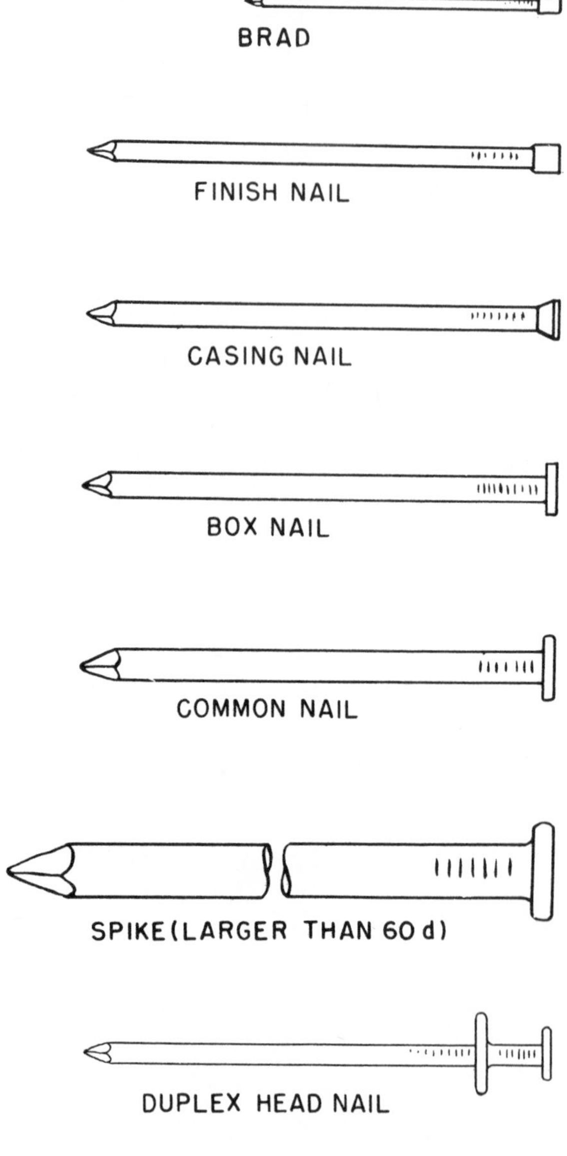


*Nail varieties.*

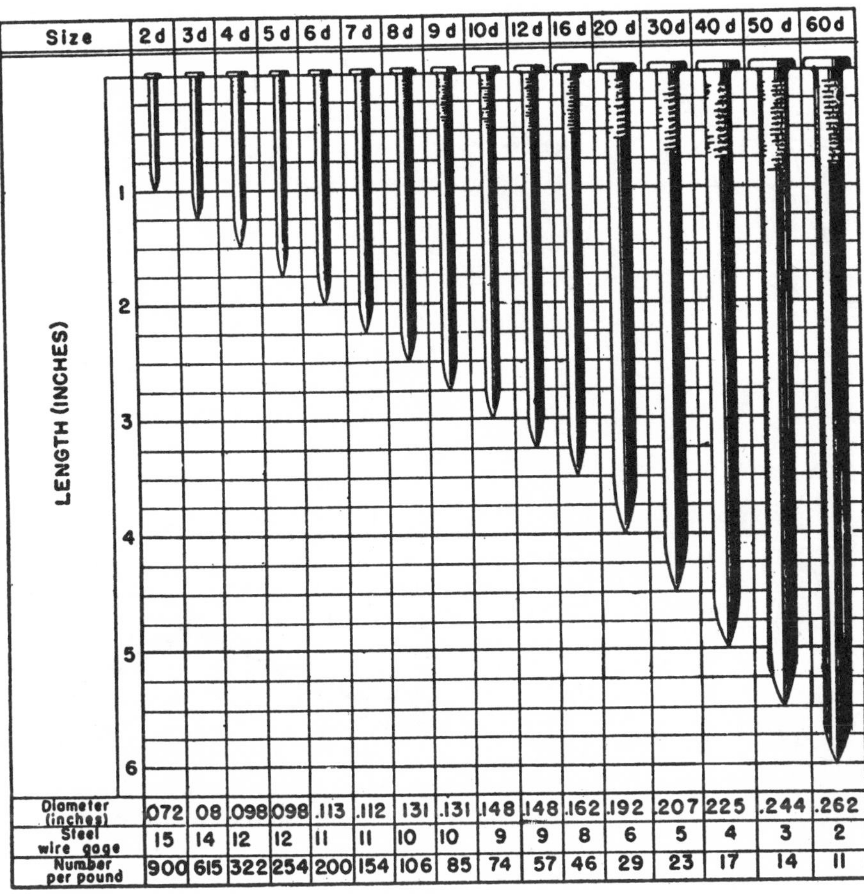


| Size | 2d | 3d | 4d | 5d | 6d | 7d | 8d | 9d | 10d | 12d | 16d | 20d | 30d | 40d | 50d | 60d |
|---|---|---|---|---|---|---|---|---|---|---|---|---|---|---|---|---|
| Diameter (inches) | 072 | 08 | .098 | 098 | .113 | .112 | 131 | .131 | .148 | .148 | .162 | .192 | .207 | 225 | .244 | .262 |
| Steel wire gage | 15 | 14 | 12 | 12 | 11 | 11 | 10 | 10 | 9 | 9 | 8 | 6 | 5 | 4 | 3 | 2 |
| Number per pound | 900 | 615 | 322 | 254 | 200 | 154 | 106 | 85 | 74 | 57 | 46 | 29 | 23 | 17 | 14 | 11 |

*Common Nail Sizes.*

for general work. *Box* nails are relatively thin and have flat heads. They were first used for nailing together boxes built of wood that was thin and therefore split easily. *Finish* and *casing* nails have small heads that can be set beneath the wood surface. They are used for furniture, cabinets, and for trim work. *Brads* are small finishing nails and are used to nail thin stock such as molding.

Use a *nail set* to drive a casing, finishing, or brad nail beneath the surface of the wood.

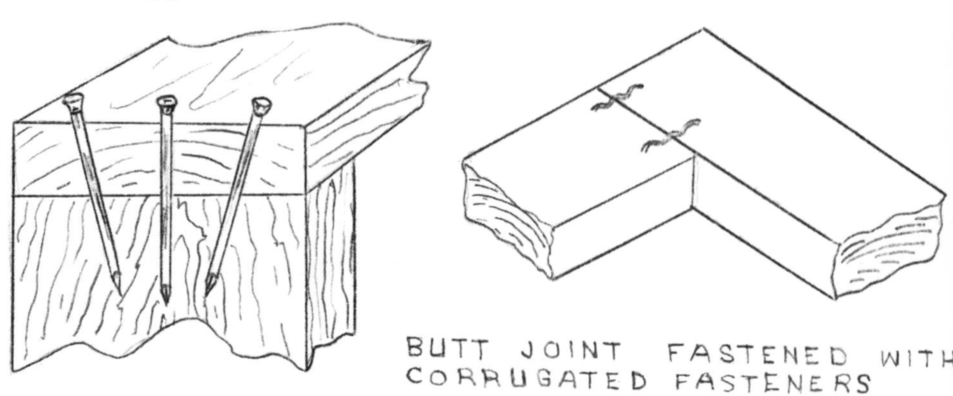

BUTT JOINT FASTENED WITH CORRUGATED FASTENERS

NAILS DRIVEN IN AT AN ANGLE

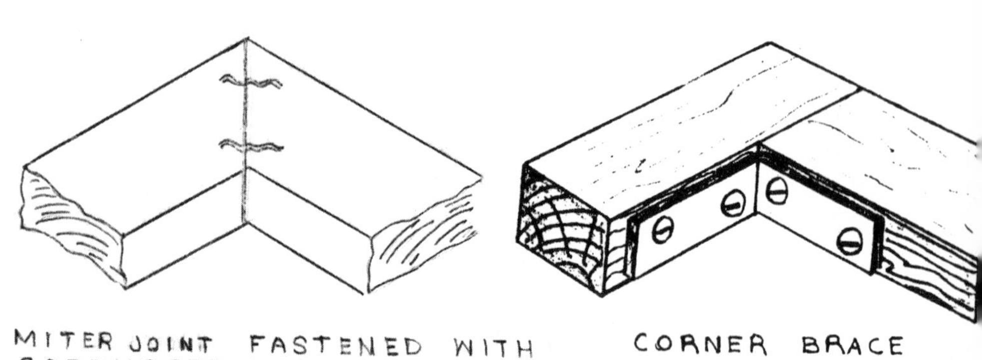

MITER JOINT FASTENED WITH CORRUGATD FASTENERS

CORNER BRACE

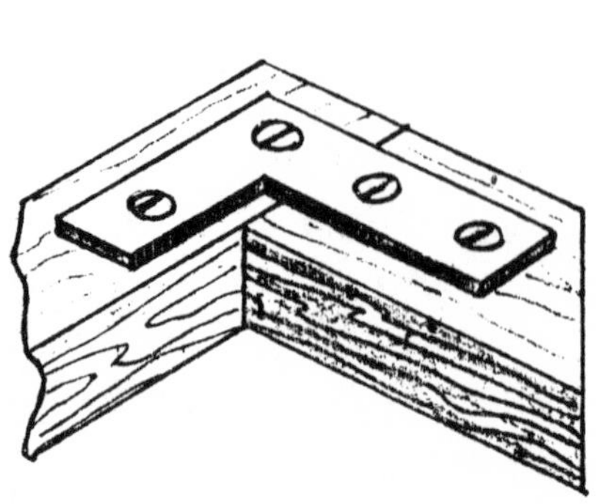

FLAT T-PLATE

FLAT CORNER PLATE

## Braces and Fasteners

Metal braces are supports used in making 90-degree angle joints. The *inside corner brace* is a strip of metal with screw holes drilled through it; it is bent to a 90-degree angle, forming an L shape, which is excellent for bracing corners. The *T plate* is nothing more than a flat drilled plate made in the shape of a T. It screws flat onto the surface of the work. The *flat corner plate* is used for the same support purposes, but it is shaped like an L. A *mending plate* has a drilled straight shape and is used for support in straight lines.

These braces are ideal for use in repair work and construction of light framed cabinets. Be sure to use the braces only where they will be concealed or where appearance does not matter.

If maximum strength is not required from a joint, a *corrugated metal fastener* can be used. Corrugated fasteners are best suited for light objects, such as screen and picture frames and simple boxes. Corrugated fasteners are sharpened on one edge for ease in driving them into the joint. Before installing the fastener into the joint, be sure that the joint is held together as tightly as possible.

## Furniture Glides and Casters

When trying to determine if your furniture project needs a glide or a caster, just ask yourself this simple question. Do I plan to move the furniture piece often or leave it stationary? If you don't plan to move it frequently, use a glide; otherwise, use a caster.

You can purchase glides in a variety of sizes with either steel or plastic bottoms. The *nail-type* glide is the most simple to install, although it is not adjustable. The screw-type glide must have a hole drilled to the diameter of the screw, and it can be adjusted to vari-

FURNITURE GLIDES.

NAIL GLIDE

NAIL GLIDE.

ADJUSTABLE SCREW GLIDE.

CASTERS.

PLATE CASTER.

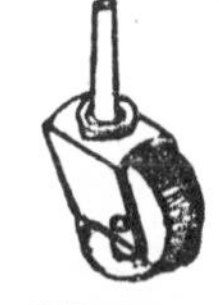

ADJUSTABLE STEM CASTER.

ous heights for an uneven floor or an uneven project base.

Casters are available in two styles: the adjustable *stem-type* caster and the non-adjustable *plate* caster. If using the stem caster, a hole must be drilled into the base of the furniture piece. The drilled hole is made to accept the sleeve, which in turn receives the stem of the caster.

The plate-type caster is simply screwed to the base bottom. This type of caster cannot be adjusted to various heights.

In order to facilitate swiveling, all casters are constructed with ball bearings. Casters are made with either plastic or rubber wheels. The rubber wheel caster is best on a hard surface such as slate, wood, or concrete. The plastic wheel caster is best used on soft surfaces such as carpet.

To stop the furniture piece from rolling use a *locking* caster. This caster locks in place by means of a small lever on the outside of the wheel.

# WOODWORKING JOINTS

There are two kinds of woodworking joints: *lay-up* and *assembly*. Lay-up joints (also called *edge joints*) are used for building up the dimension of stock. Assembly joints are those used in assembling members which have been cut to a certain shape and dimension.

The joint's purpose is to provide strength without detracting from appearance. Most joints are held together by the use of fasteners such as nails, screws, corrugated fasteners, wedges, or pins. Glue is used as an adhesive when a metal or wood fastener is not desired.

The strength of a woodworking joint depends largely on the accuracy of the fit and the quality of workmanship used in applying the fasteners or adhesive. When clamping with several bar clamps, it is advisable to place the clamps alternately, one up and the next one down, which prevents the surface from buckling. Clamps should be placed approximately 18 to 24 inches apart.

## Lay-up or Edge Joints

*(used for joining boards)*

### Butt Joint

The butt joint is the most frequently used joint. It is constructed by butting together one member (end or edge) of wood to the end, edge, or surface of another. It is secured by using glue, nails, or screws. It is the weakest kind of joint and should not be used

where a great deal of pressure will be applied. The dowel, spline, or half-lap joint are recommended.

### Dowel Joint

Dowels are used to give additional strength to a butt joint by adding resistance to cross-strain. Drill holes for the dowel pins in the edges of the pieces to

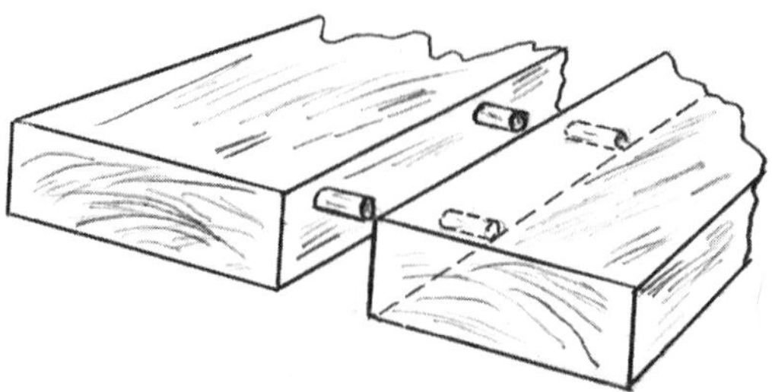

be joined. For accuracy use a doweling jig. Apply glue to all parts, insert the dowels in one piece, continue to line up the other pieces, and join them together using clamps.

## Spline Joint

The spline joint is made by cutting matching grooves or keyways into the edges of two boards. The spline is a narrow strip of thin wood made to fit the dimensions of the groove of both boards. The groove

can be cut to any depth; the width is usually one-third the thickness of the stock. Cut the spline to fit into the grooves. Apply the glue to the parts to be joined. Install the spline into the matching grooves and fasten the boards together using clamps.

## Rabbet or Half-Lap Joint

Two pieces can be fastened together by the rabbet or half-lap joint. The joint is made by cutting two rabbets one-half the thickness of the stock, hence a one-

inch board will have a rabbet cut to one-half inch thickness. When the two boards are fastened together with glue, they will equal a board one-inch thick.

Cut the rabbet joints to the necessary thickness: check them for proper fitness by fitting them together before gluing. Apply glue and clamp.

## Tongue-and-Groove Joint

The tongue-and-groove joint is made by cutting two rabbets of equal dimensions on one board's edge

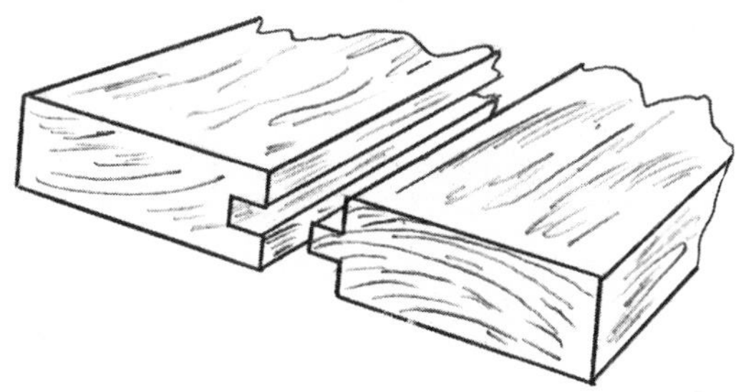

to achieve the tongue, then by cutting a groove to the width and depth of the tongue on the matching board. This is fastened together by applying glue and clamps. The tongue and groove opening should not exceed half the thickness of the stock.

# ASSEMBLY JOINTS

*(used for framing cut members together)*

## Butt Joint

The simple butt joint is the weakest joint used in assembling cut parts. The butt joint is constructed by butting one piece of wood to another at a right angle. The surface contact made is with the end grain of the

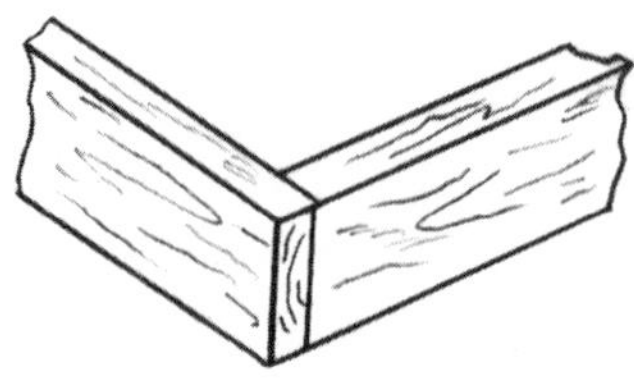

wood. Because the end grain is extremely porous and may be difficult to glue, the butt joint should be reinforced with glue blocks screwed to the inside for added strength.

## Dowel Joint

The dowel joint gives added strength to the butt joint by adding resistance to the cross-strain of the stock. Drill holes for the dowel pins in the edges of the

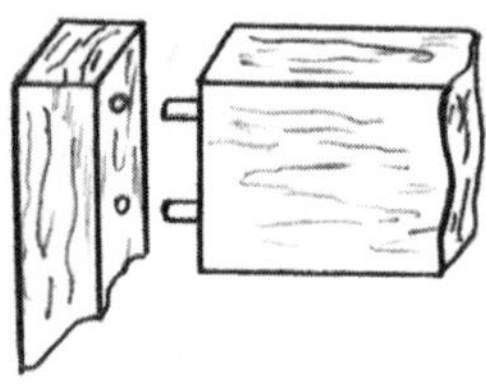

two pieces to be joined. Use a doweling jig for accuracy. Apply glue to the dowel pins and board edges. Insert the dowels in one piece, and continue to line up the piece. Join them together using clamps.

**Miter Joint**

A miter joint is a butt joint in which two pieces of stock are cut at the same angle, usually 45 degrees, in order to form a 90-degree angle. Because the joined

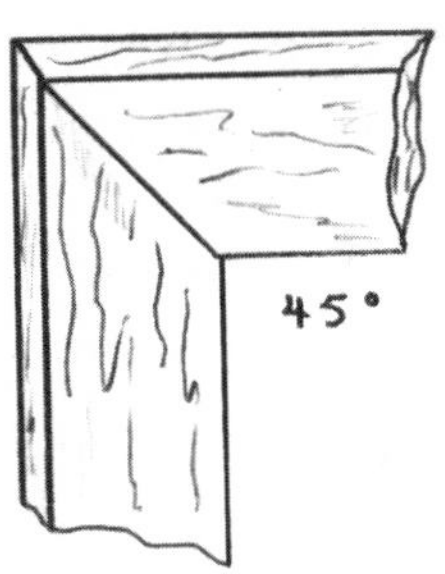


surfaces are end grains, the joint is weak and should be reinforced with a spline, dowels, or corrugated fasteners.

**Half-Lap Joint**

This joint has remarkable strength and is therefore one of the most frequently used right-angled joints. It is

constructed by cutting away half the thickness of each member to be joined so that when fastened together its thickness equals that of one member.

**Rabbet Joint**

A rabbet joint is a recess cut along the end or edge of a board. It is usually used in panel and drawer con-

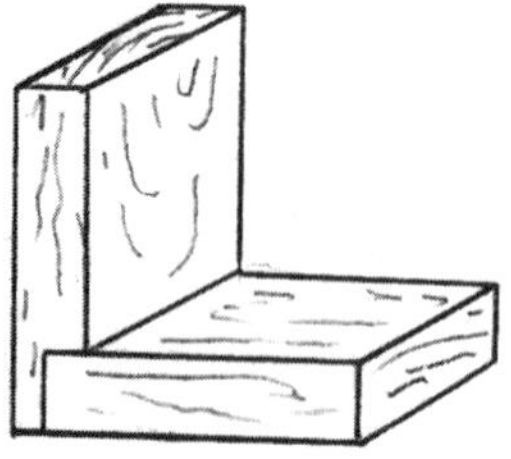

struction. The cut recess should be half to two-thirds the thickness of the stock. The rabbet joint should form a 90-degree angle.

### Dado Joint

A dado joint is often confused with a groove joint. A dado is a recess cut *across* the grain of wood into which another board is to fit. A groove is a recess cut *with* the grain of wood.

Dado and groove joints should never exceed more than one-half to two-thirds the thickness of the stock in which they are cut. The joint should fit snugly and the depth of the cut must be even.

## Gluing

Today the variety of glues available to the consumer seems almost endless. Therefore I will deal only with the most commonly used glues on the market.

### Polyvinyl Acetate (PVA)

This is an excellent glue for most household projects, such as ceramic work, craftwork, and furniture construction. PVA glue will give an excellent bond but cannot withstand excessive moisture or high temperature. This glue should set for at least 8 hours in a 65 to 70-degree F. temperature.

### Polyvinyl Chloride (PVC)

This is an excellent waterproof adhesive used on wood, metal, china, glass, or porcelain. The drying time varies, therefore brand specifications should be followed.

### Cellulose Glue

This is best suited for small repairs to furniture and model work, and is superior for joining chinaware and glass. It also can be used on most fabrics. It gives a fast-drying, colorless joint. The bond strength is increased by applying two coats to both surfaces.

### Epoxy Glue

This glue is excellent for joining glass and metals as well as for china repairs. Epoxy adhesives are waterproof and oil-resistant. They are excellent for bonding dissimilar materials, such as plastic to metal, glass to metal, glass to concrete, etc. Epoxy glues are made in regular and fast-setting types. They are "two-part" glues: a resin and a hardener are mixed in equal amounts.

### Liquid Hide Glue

This is the traditional glue used by cabinetmakers. Because it makes an excellent bond that holds up well under a heavy load, hide glue is best suited for furniture repair and construction. Do not use hide glue on projects that will be exposed to water or high humidity.

## Joint Bonding

Joint preparation is extremely important, for even the thinnest film of dust, dirt, or oil will prevent an adhesive from bonding correctly. To bond wood joints, be certain the surfaces are square and are sanded properly; all sanding dust must be cleaned from the surfaces before applying the glue.

It is important that the joints be fitted tightly with the aid of clamps such as bar clamps, screw clamps, or spring clamps. Repairs made to china, glass, and plastic are clamped with makeshift devices such as small jigs or with masking tape or elastic bands.

If a wood joint is loose, it can be tightened by filling the space with small wood slivers. Cut the slivers to the size needed, then apply glue to them. Press each sliver into the gap. For smaller jobs use crack filling adhesives, such as plastic wood.

## Finishing the Project

It is crucial that the project be carefully sanded to a smooth surface before any application of finish. Although power sanders may help to do much of the sanding, only careful hand-sanding will yield a professional result. It is foolish to think that stain or paint will conceal a poorly prepared surface. On the contrary, they tend to draw attention to a poorly prepared surface.

Sanding of the project should be done in stages starting with 60 to 80 grit paper for rough work, then continuing with 120 to 150 grit paper for smoothing the project. It is good practice to do all sanding in a straight line with the grain of the wood.

Splits, checks, or nail holes should be filled with plastic wood or stick shellac. These products can be purchased in a multitude of colors to match the color of your wood or stain.

A finish should not be applied under extreme conditions of dampness or coldness.

When staining the project, apply the stain in an area that will receive the same amount of light that the project will receive while in use.

The purpose of stain is twofold. It is translucent, which allows it to emphasize the beauty of the grain, and it allows the craftsman to color the project as desired.

There are several types of stains, oil stains and dye stains being most common. They are applied in a liquid state and require varying amounts of drying time. When the stain has been thoroughly dried, the surface should be given several coats of clear varnish, linseed oil, or lacquer for protection.

**Oil Stains**

These are composed of finely ground powders mixed to a paste with benzene or turpentine. The stain can be applied with a rag or brush and can be rubbed on in any direction. Oil stains should be given 15 to 20 minutes to dry, after which it should be rubbed off in the direction of the grain with a clean cloth. If the color tone is too light, allow the first coat to dry overnight, then apply a second coat.

**Dye Stains**

Dye stains normally have a water or alcohol base. These stains soak into the wood very fast: therefore, they should be applied in very light coats. Stains of this type are known for their richness of color and their ability to bring out the grain of the wood.

Water-base dye stains must be mixed, for they come in powdered form. They can be applied by spraying, brushing, sponging, or with a rag. The powder is mixed with hot water. The advantage of mixing your own stain is that you can control the color: if the color is too light, it can be darkened by adding more powder;

dark colors can be made lighter by adding more water. Two coats of dye stain should be applied for best results.

A disadvantage of water-base stain is that it raises the grain of the wood. To compensate for this it is best to allow the stain to dry overnight, then sand the surface lightly with 280 grit sandpaper.

Alcohol-base dye stains have the advantage of not raising the grain of wood and of being sold premixed in a variety of colors. Their color intensity can be reduced by adding alcohol. Stains with an alcohol base can be applied to the wood with a rag, with a brush, or by spraying.

These safety precautions should be followed:

1. Wear gloves when applying.
2. Do not work near an open flame or high temperature.
3. Do not smoke while applying the stain.

### Varnish

There are many grades and colors of varnish available, and they are sold ready to use. The better grades will expand and contract with the wood to which they are applied, without cracking. The colors range from clear to dark brown. Varnish will dry to form a scratch-resistant surface that is hard enough to rub down between applications.

Cellulose varnish, better known as polyurethane, is extremely durable and is unaffected by water. It does not discolor when exposed to light.

Spar varnish is available with chemicals added to make it resistant to salt water.

Turpentine acts as a suitable thinner and cleaning agent for all varnishes.

Before applying varnish to bare wood, the pores should be filled and the surface should be as free of dust as possible. At least two coats of varnish should be applied; the rubbing between each coat should be done with fine steel wool or extra fine sandpaper. As you rub between coats, you will feel the varnish finish smooth out as the bubbles and dust particles flatten. It is wise to allow plenty of drying time for a hard finish coat. The harder the varnish surface becomes, the easier it is to rub down between coats.

Varnish is sold as high gloss, satin finish, or flat finish.

### Lacquer

Lacquer is an excellent finish, for it tends to dry fast and does not conceal the wood color. It can be identified faster than other finishes because of its banana-like odor.

Lacquers are available to be brushed on or sprayed. Lacquer that is brushed on takes longer to dry than sprayed lacquer. Brush-type lacquer has a heavier consistency than the spray-type and therefore requires a longer drying period. When thin coats are applied, the drying time is always reduced.

Once the work is sanded to a smooth surface and stained, apply the first coat of lacquer by brushing or spraying. A good spray job should have at least three to four light coats. Rubbing between each coat is not essential because lacquer tends to leave a smooth surface.

Care must be taken when using a spray gun because heavy spraying has a tendency to run, and it is almost impossible to lift the runs. Lacquer should be thinned with lacquer thinner *only*.

To brush on lacquer, apply the first coat as evenly as possible in the direction of the wood grain. After an overnight drying period, apply the second coat. After the final application has dried for 36 hours, you may find some imperfections in the form of uneven surface, dust particles, and small pits. These can be eliminated easily by rubbing the surface down with wax applied to a steel wool pad.

All safety rules for applying a combustible finish should be followed carefully when working with lacquer.

### Finishing Oil

Applying oil to wood is one of the oldest and best known methods of wood finishing. It is easy to apply and to maintain. A wood surface finished with oil will tend to dry out in time and can be easily revived with one new application of oil. If the surface is dirty, all that is required is to wipe it down with turpentine or rub the dirt spots with steel wool.

Linseed oil is probably the best type of oil finish, for it tends to last longest. When working with linseed oil, one should use boiled linseed oil diluted with one part of turpentine for each part linseed oil. The formula may be adjusted to suit individual taste.

The project should be sanded to a smooth surface, with dust thoroughly removed. Wipe or brush the oil on generously with a clean cloth. Allow the oil to saturate the wood, and if dull spots appear, add more oil. When the wood can absorb no more oil, wipe off the excess with a clean cloth. Rub the surface down with a lint-free cloth folded into a pad.

To work with oil successfully one must rub hard enough with the heel of his hand to produce heat. After allowing the oil a drying period of 24 hours, repeat the above procedure 4 to 6 times to produce a beautiful satin finish.

### Pumice

Pumice is a light, porous material obtained from volcanic lava. It acts as a sharp cutting material which smooths and polishes by producing fine, hairlike scratches. Pumice is available in powdered form and is used with oil, water, or paraffin to produce a smooth finished surface.

A light film of oil or lubricant is applied to a cloth

pad. Pumice is sprinkled over it. Rub the pumice in the direction of the grain until you achieve a smooth finished surface.

### Wax

Paste wax adds protection and life to a finished surface by covering it with a tough film. The film protects the surface by making it water-repellent and abrasion-resistant.

Some waxes require a damp cloth as an applicator. When applying the wax, one should work in an area of approximately 3 square feet at a time. The paste should be applied in an even layer. After its recommended drying time, it should be buffed thoroughly. Polish the adjacent area, making sure to blend the entire finish until no bare spots remain.

## How to Use the Plans

As you look through this book, you'll notice that most of the drawings are *working* or *view* drawings. For each project the working drawings give one, two, or three views, typically front, top, and end. Some of the project drawings are pictorial drawings; they show how the project will look when fully constructed.

Before beginning a woodworking craft project, it is important to study the working drawing, which indicates the size and shape of all parts. The working drawing will serve as your blueprint. By studying it, you can determine the required thickness, width, and length of each part required for construction of the project.

The working drawing also illustrates the shape of curved and scrolled parts by graphing the scrolled or curved part in squares of specific size. In order to reproduce the design on your wood, merely lay out the required grid or squares on the stock, then draw the suggested shape square by square (see illustration).

A suggested list of materials is given for each woodworking project. Exact dimensions are given for all materials required. Helpful hints advise you where to locate hard-to-find materials.

# CABINET AND FURNITURE CONSTRUCTION HINTS

### Box Construction

Box construction, the most simple of all cabinet and furniture fabrication, is used to build drawers, chests, and numerous types of boxes (e.g., toy boxes). The box is a simple four-sided structure with a bottom and often a top lid. The boards used in box construction are always assembled so that the grain runs in the same direction on all four sides. The corners may be joined with simple butt or rabbet joints or with more elaborate joints. The bottom of the box is usually set into a groove or rabbet cut.

### Case Construction

Kitchen cabinets, stereo cabinets, bookcases, and the like are examples of case construction. Case construction is simply a box turned on its side. Therefore, its method of construction is similar to that used in box construction. The corners are joined with a miter, rabbet, or similar joint. A rabbet is cut around the rear edges so that the back, which is most often plywood or hardboard, will fit snugly. In case construction, the same material serves as both the interior and exterior of the assembled structure. If the structure is designed to have fixed shelves, the shelves are usually installed into dado joints, although they can also be held in place with nails and screws. Movable shelves are usually mounted on adjustable shelf brackets, which can be purchased in various lengths and styles. The adjustable track can be installed flat onto the surface, or it can be recessed into a groove made to size. Plywood and other types of sheet material are well suited for case construction. If plywood is used, it is advisable to face or nose the exposed front edge.

### Frame Construction

Frame and case construction appear to be similar, but the fabrication techniques differ. In frame construction the material used to cover the frame is usually lighter in weight. Instead of using ¾-inch-thick solid wood for the cover, ½-inch-thick plywood or particle board is used. Prefabricated bathroom vanities and kitchen cabinets are usually constructed from frames. The cabinet frames are most like to be ¾-inch thick by 1½ to 2 inches wide. The front stiles (vertical members) of the frame usually overlap the sides in order to conceal the cheaper edges of the frame cover material.

### Leg and Rail Construction

In leg and rail construction only four rails and four legs are usually used. The joints commonly used to join the corners are the mortise and tenon and the dowel joint reinforced by corner glue blocks or various metal hardwares. Of these two the dowel joint is the more simple to construct. The only difficulty with the dowel joint lies in keeping the holes completely aligned. A good adhesive must be used for maximum strength.

To secure a top to the leg and rail assembly, cleats or special metal hardware may be used.

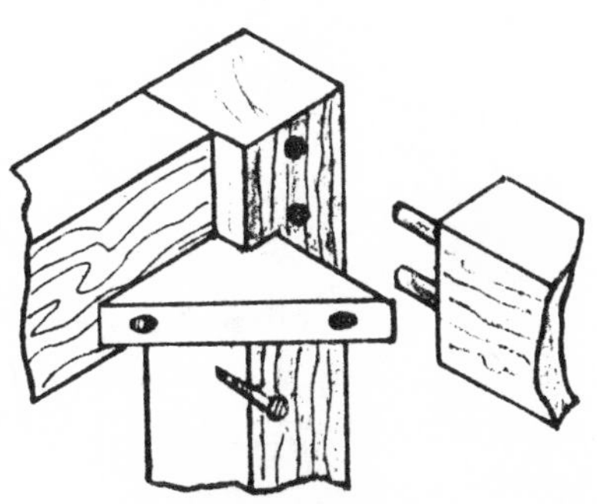

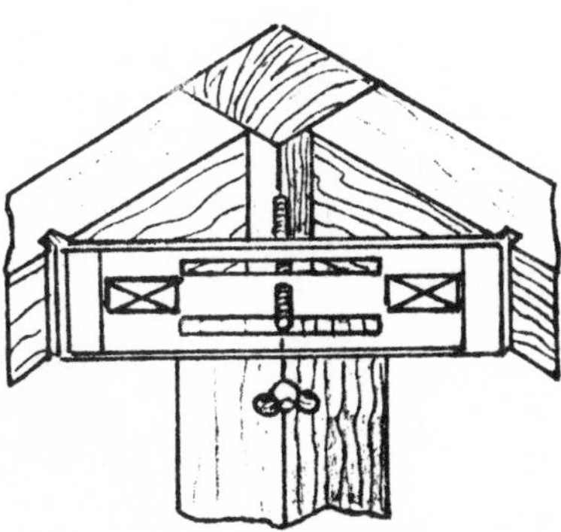

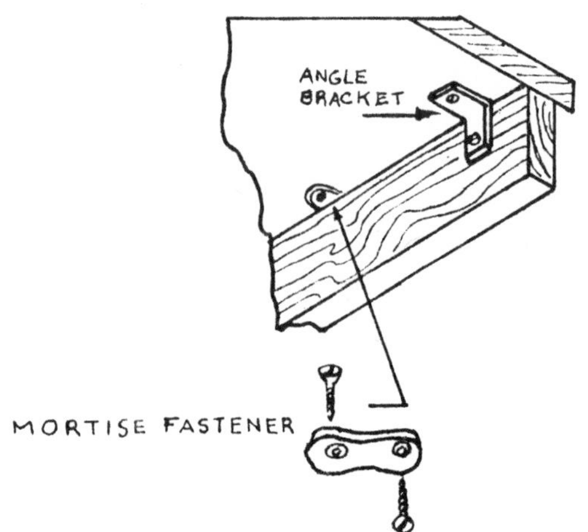


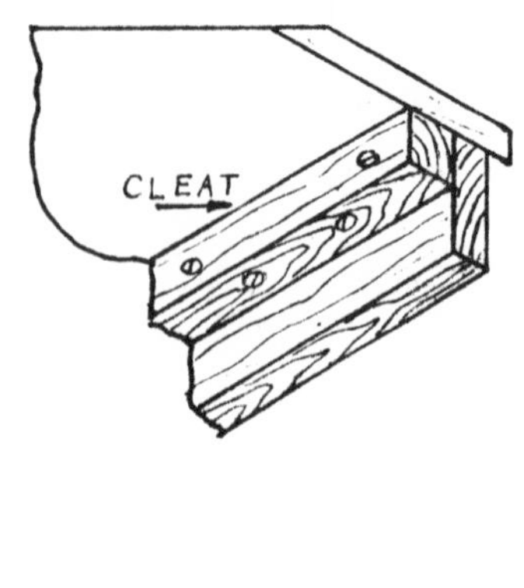


METHODS OF REINFORCING TABLE TOPS AND RAILS TO LEGS.

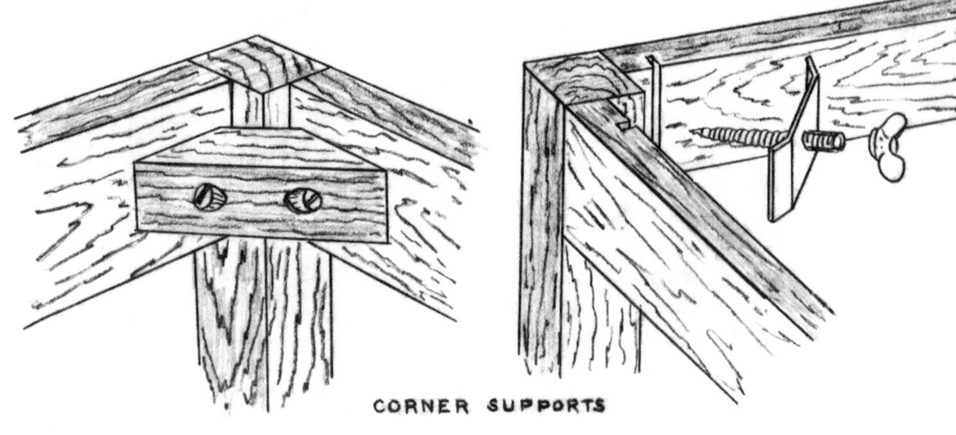

CORNER SUPPORTS

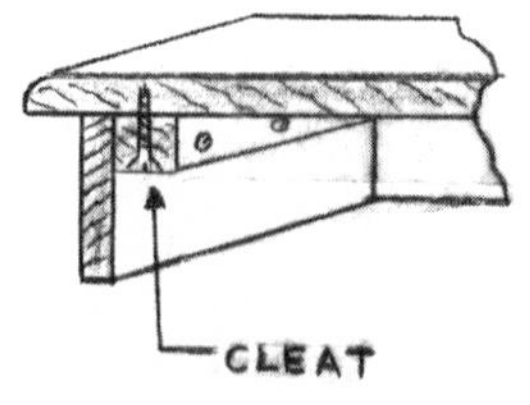


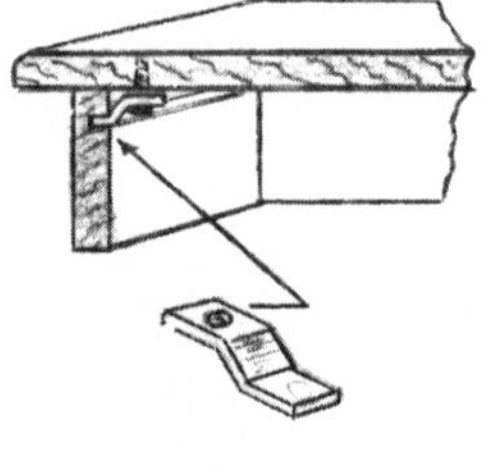

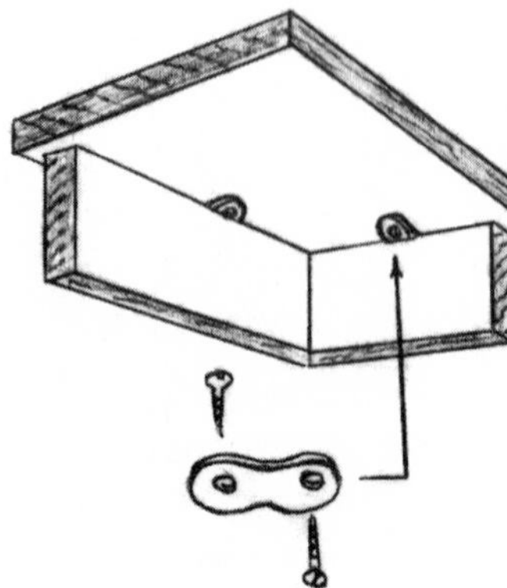

METAL FASTENERS

## How to Construct Frame Panels

Most frame panels are used in constructing doors for cabinets and furniture. The frame surrounding the panel consists of two vertical side members, called stiles, and two horizontal top and bottom members, called rails.

The stiles and rails are always constructed of solid wood, and they range from ¾-inch to 1-inch thick by 1½ to 2½ inches wide.

There are numerous methods for securing the corners of the stiles to the corners of the rails. These include the mortise and tenon joint, the dowel joint, the lap joint, and the simple butt joint and miter joint. The panel, which can be solid stock or standard sheet stock such as hardwood or plywood, is fastened into a groove cut into the frame. The panel can also be placed in a rabbet instead of a groove. The rabbet cut is always used for placement of glass, mirrors, or plexiglass.

Some fine furniture often has raised panels or various decorative cutouts and carvings in the panels and rails.

## Constructing a Raised Panel

In making a raised panel, it is important to use only solid wood stock glued up to the correct width. Adjust the tilting arbor of a table saw to an approximately 10-degree to 12-degree angle with the depth of cut ranging from 1 inch to 3½ inches. The table saw rip fence should be placed approximately 3⁄16 inch to ¼ inch away from the highest point of the tilted blade. Using the 10-degree to 12-degree angled setting, proceed to make a cut on all four edges.

Return the tilted blade to a 90-degree angle and adjust the depth of cut to approximately ⅛ inch. Set the rip fence so that the saw cut will intersect the first angled cut at its full depth. Proceed to cut all four edges in completing the camfered cut. Set the panel into a groove or rabbet that has been precut in the frame.

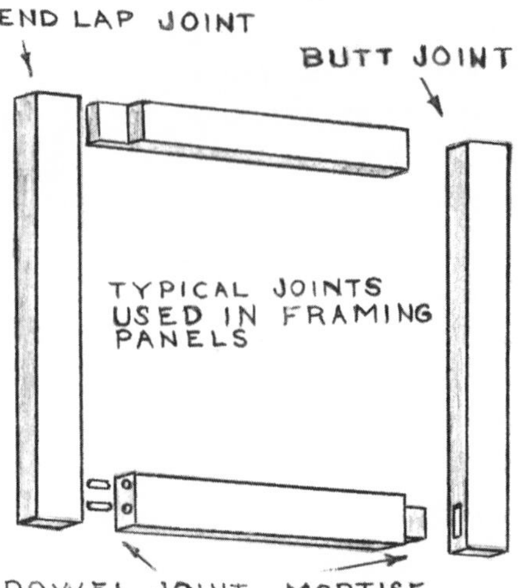


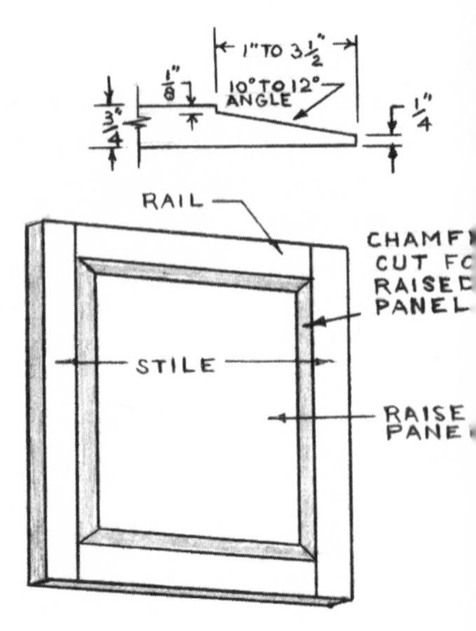


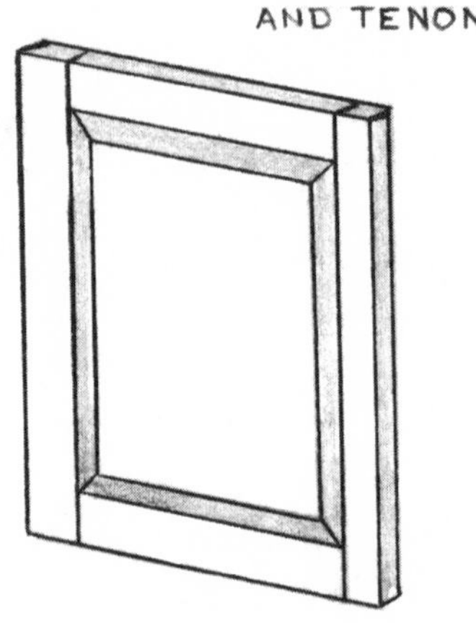

RAISED PANEL FRAME

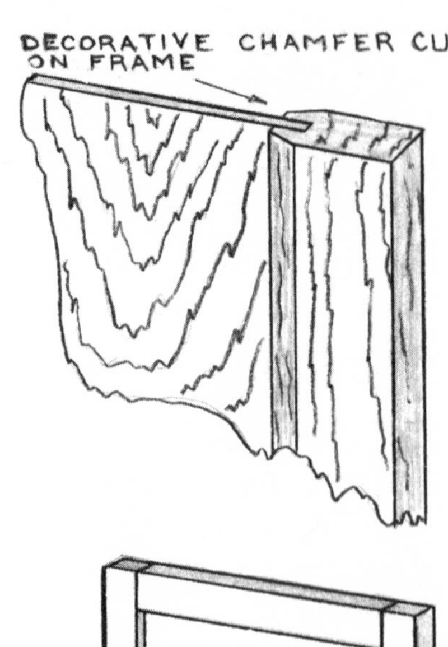

DECORATIVE CUT IN RAIL

PLAIN PANEL FRAME

*Note:* If a plain or raised panel is set into a frame with a rabbet cut, it should be held in place with a ¾-inch half-round molding lapped over both frame and panel.

## How to Construct Drawers

There are two basic drawers most often used in furniture and cabinet construction: the flush front drawer and the lip front drawer. The lip front drawer has an overlap of ⅜ inch on all sides, thereby partially covering the cabinet frame, whereas the flush drawer fits flush into the front of the cabinet frame.

Simple box construction, described above, is used for the fabrication of drawers. Simply put, drawers consist of two sides, a front, a back, and bottom. Corner joints can consist of the simple butt joint, requiring simple tools, but quality drawers may have the difficult dovetail joint, requiring more time and the use of electric power tools. The rabbet joint can be rated medium-difficult. The bottom of the drawer is usually made from ¼-inch to ⅜-inch plywood which is fitted into a groove placed approximately ½ inch above the bottom edge of the sides and back.

One should practice care in assembling a drawer. Glue and nails should be used to fasten everything but the bottom, which is left free in its groove to allow for swelling and shrinking of the drawer. It is essential for the drawer to be perfectly square or it will not fit into the cabinet opening.

A drawer should be constructed only after the furniture or cabinet in which it is to be fit has been built. Start by determining the type of drawer front desired: flush front or lip front. Be sure to match the grain and color of the wood if a cabinet will contain more than one drawer.

The front of a drawer is usually ¾-inch thick; the sides and back are usually ½-inch thick. Try to select

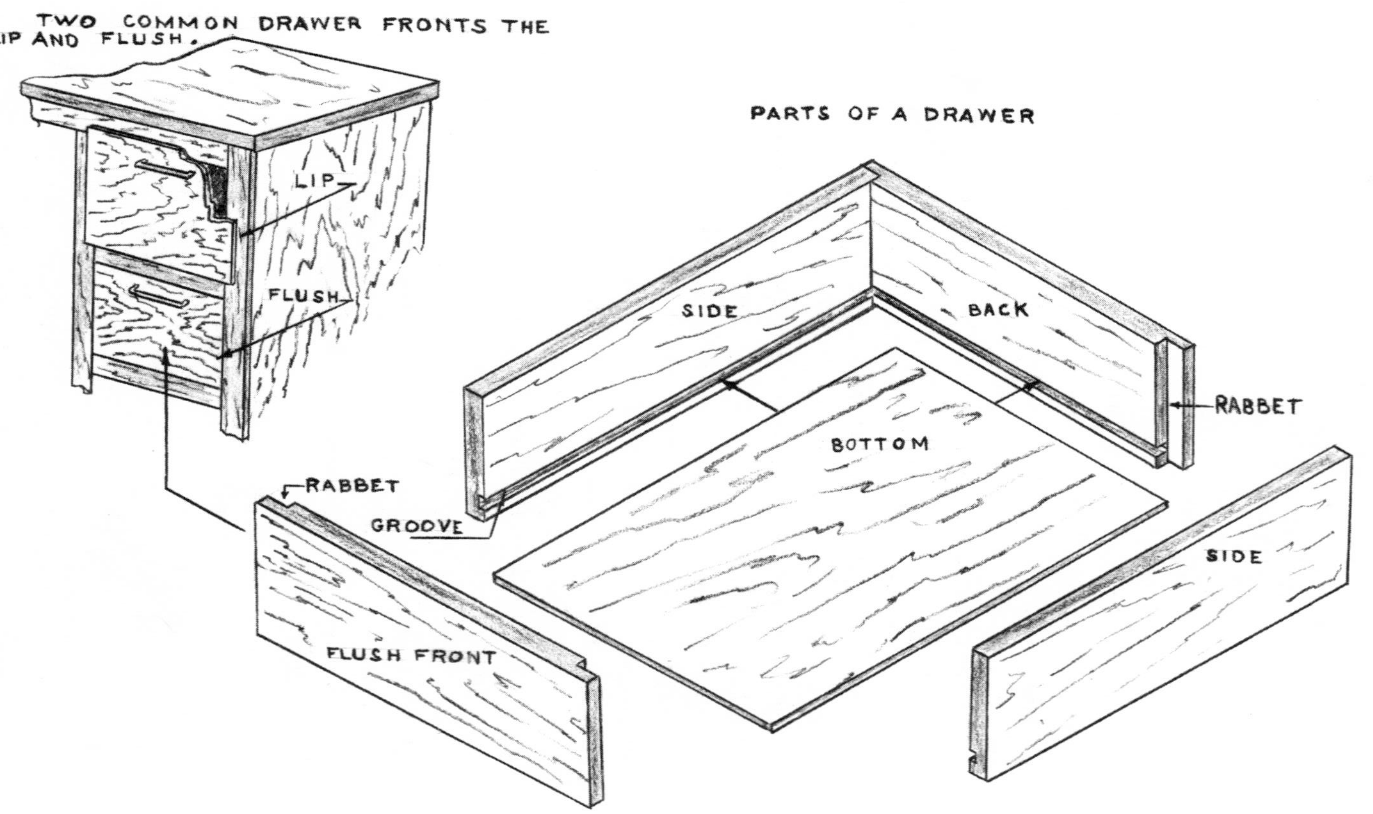
TWO COMMON DRAWER FRONTS THE
LIP AND FLUSH.
LIP
FLUSH
PARTS OF A DRAWER
SIDE
BACK
RABBET
BOTTOM
RABBET
GROOVE
SIDE
FLUSH FRONT

knot-free stock that is not warped. Pine and oak are good woods to use. A ⅜-inch overlap drawer is safest to use, for it will cover the drawer opening even if it shrinks. A ⅜-inch lip drawer front should be ¾ inch longer and ¾ inch wider than the cabinet opening. A flush drawer front should be 1/16 inch narrower and ⅛ inch less wide than the cabinet drawer opening.

## How to Construct Drawer Guides

Most of the projects in this book require drawer guides. Several styles of wooden guides can be made, and they should be chosen based on the size drawer with which they are to be used. Small drawers do not need guides, or they may just need a simple strip glued to each side. Large drawers, which are designed to carry more weight, require stronger guides.

The single center guide is nothing more than a simple ½-inch by 1-inch piece of stock fastened in the center of a dust panel or drawer divider board located under the drawer bottom. Glued on the drawer bottom are two ½-inch by 1-inch guide rides which override the center guide placed on the dust panel. It is necessary to cut a notch in the back drawer panel to match the opening created by the two guide rides. A single center guide can be made relatively easily by using a ½-inch by 1-inch notch cut in the center back drawer panel to track on the center guide. A top guide is used to prevent the drawer from tilting down when the drawer is opened.

Side guides are fastened to the lower corners of larger drawer openings. In using this guide technique, it is necessary to use a top guide to prevent the drawer from tilting down when opened.

If a drawer is to be placed directly under a table top, the drawer side guides are constructed in the shape of an L. The L-shaped guides are secured under the table top, and guide rider strips are fastened flush to the top edge of the drawer sides.

*Note:* It may occasionally be necessary to use soap

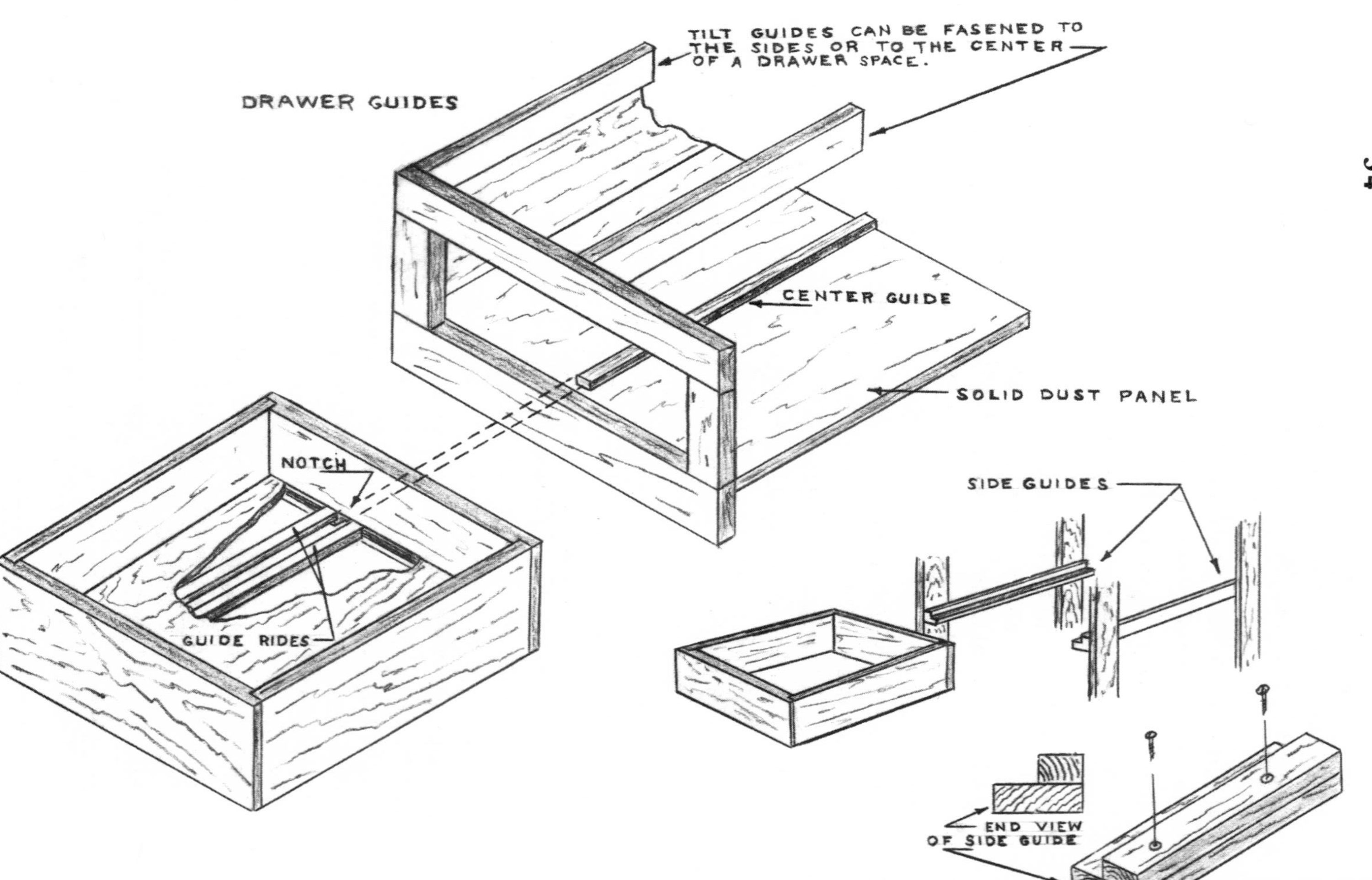
DRAWER GUIDES
TILT GUIDES CAN BE FASENED TO THE SIDES OR TO THE CENTER OF A DRAWER SPACE.
CENTER GUIDE
SOLID DUST PANEL
NOTCH
GUIDE RIDES
SIDE GUIDES
END VIEW OF SIDE GUIDE

or wax on the guides and riders to assure smooth opening and closing of the drawer.

### Commercial Drawer Guides

Many drawer guides are made commercially. Perhaps the best guide on the market is the roller type coupled with bearings, for it will withstand the heavy load commonly found in dresser drawers, file drawers, and kitchen cabinet drawers.

### How to Construct Doors

Like drawers, most doors made for furniture or cabinets are of flush or lip construction. The material used for doors is usually plywood, solid stock, or frame and panel.

Lip doors are designed to set over the frame of the cabinet in the same way that drawer fronts overlap. Each door is always cut ¾ inch longer and wider than the cabinet opening to allow for the ⅜-inch lip. On the inner edge of the door is a ⅜-inch-wide and ⅜-inch-deep cut made completely around to allow the door to fit over the frame. The lip door is attached to the frame with a semiconcealed ⅜-inch offset hinge.

A flush door is fitted to the cabinet opening in the same manner the flush front drawer is fitted to its opening. The door should be approximately 1/16 inch narrower and ⅛ inch less wide than the cabinet door opening. When installing a flush door, either butt or surface hinges should be used.

## PROCEDURES FOR USING THE LATHE

### Faceplate Turning

In faceplate turning the stock is secured to a flat metal plate which is fastened to the spindle of the head stock of the lathe. As the lathe revolves, the wood fas-

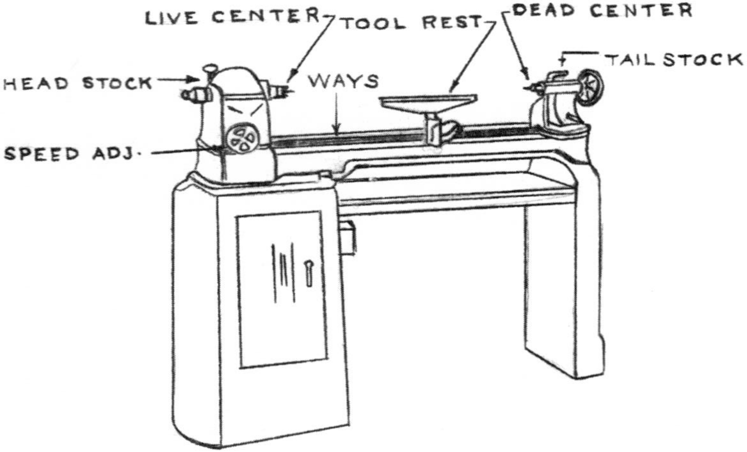

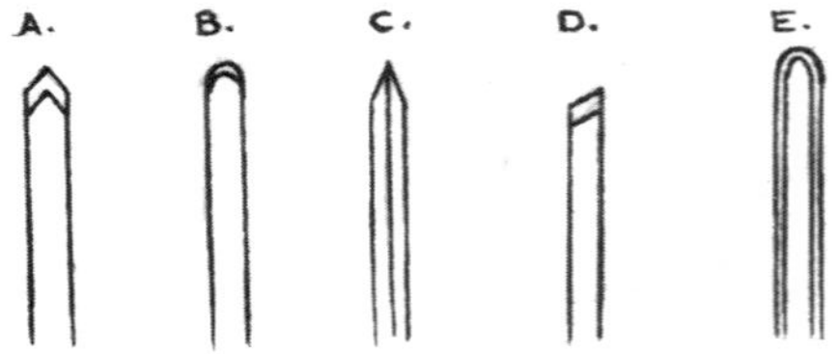

LIVE CENTER
TOOL REST
DEAD CENTER
HEAD STOCK
WAYS
TAIL STOCK
SPEED ADJ.

A. B. C. D. E.

COMMON WOOD TURNING TOOLS.
A. DIAMOND POINT. B. ROUNDNOSE.
C. PARTING TOOL. D. SKEW CHISEL.
E. GOUGE.

STEP 3.
STOCK FASTENED BETWEEN CENTERS.

STEP 1.
LAYING OUT DIAGONAL LINES TO

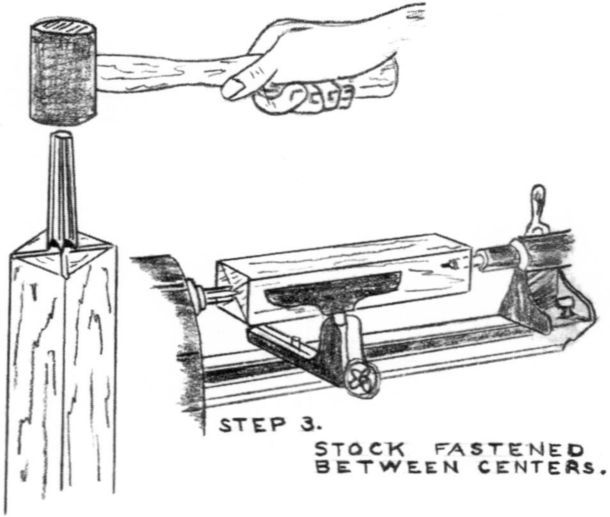

STEP 2.
DRIVING THE LIVE

STEP 1.

STOCK CUT SQUARE AND MARKED FOR FACEPLATE TURNING.

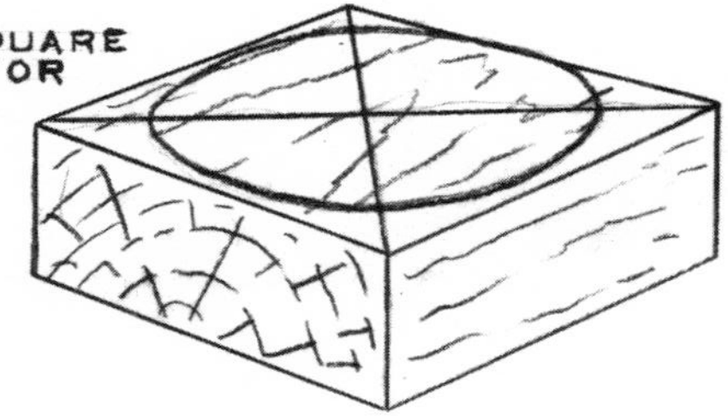

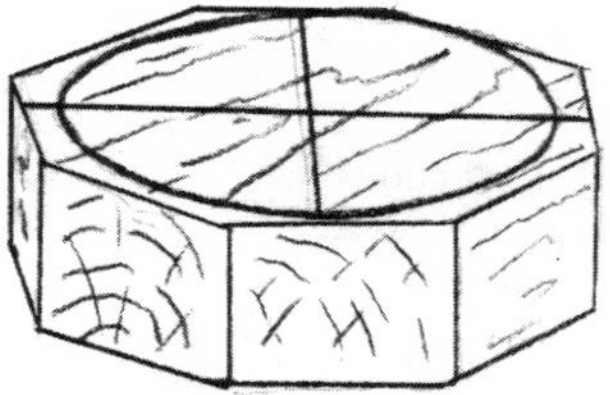

STEP 2.

CORNERS CUT FOR FACEPLATE TURNING

STEP 3.

FACEPLATE SECURED TO WOOD.

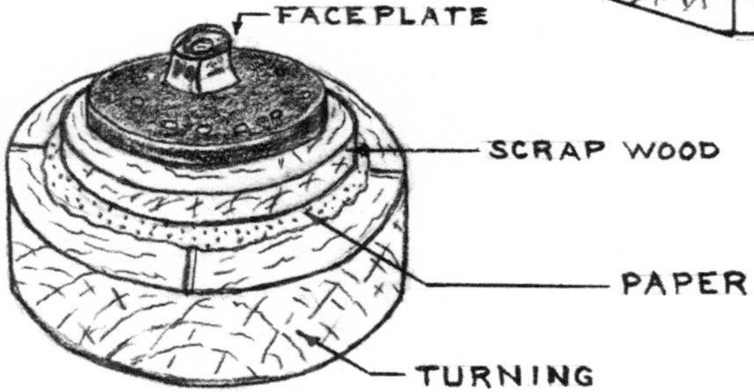

FACEPLATE

SCRAP WOOD

PAPER

TURNING

tened to the faceplate is shaped by a series of scraping and shearing cuts. The cutting tools used are the *round-nose* chisel for concave cuts, the *skew* chisel for convex cuts, the *diamond* point for V-cuts, the *parting* chisel for depth cuts, and the *gouge* for rough turning.

1. Use a band saw to cut the stock approximately ½ inch larger than the finished dimension. The corners of the stock can be cut off, or it can be band-sawed into a disk.
2. Choose a faceplate smaller in diameter than the stock to be turned.
3. Screw the faceplate to the bottom of the stock. If screw holes are not desired or there is a chance that the turning tools will come in contact with the screws while cutting, use a ¾-inch block of scrap wood glued to the base of the turning. Before gluing, insert a piece of heavy paper between the glued surfaces in order to separate the turning from the scrap wood without damaging the base of the turning.
4. Fasten the faceplate to the stock base, making sure the centers are aligned.
5. After removing the live center from the lathe, screw the faceplate to the spindle.
6. Position the tool rest so it is ⅛ inch below the center and ⅛ inch to ¼ inch away from the stock being turned.
7. Revolve the stock by hand to check ⅛-inch clearance at all points.
8. To adjust the correct speed use this rule of thumb: Pieces that are 3 inches in diameter or more should have a r.p.m. of 600; pieces less than 3 inches in diameter can run approximately 1200 r.p.m.
9. Smooth the face of stock with a gouge.
10. Reset the tool rest parallel with the ways of the lathe bed. Set it approximately ⅛ inch to ¼

inch from the outer edge and ⅛ inch below the center.

11. Check the block by revolving it by hand for proper clearance.
12. True the edge with a gouge or with a round-nose tool.
13. Complete the turning to the design of your choice or to the design on a drawing.
14. Sand the turning smooth while it is revolving on the lathe.
15. Clean the project and the lathe of dust to prevent dust particles from landing on the finish.
16. Make a pad of clean cotton cloth.
17. Moisten the pad with clear oil, or simply use beeswax.
18. Turn the lathe on to 1200 r.p.m.
19. Hold the pad against the revolving surfaces of the turning with an even pressure. Hold the finishing pad under the turning when applying the finish; this will enable you to see how smoothly the finish is being applied.
20. Apply several coats of finish until you reach your desired finish.

## Between-Center Turning

It is important to have the stock for between-center turning absolutely centered for the proper balance and an even turning of your project.

1. Select the stock. Cut to size with an allowance of approximately 1 inch for waste in length and ¼ inch for waste in turning down to the finished diameter.
2. With a straight edge intersect diagonal lines across both ends of stock to locate the centers.
3. With a back saw cut a ⅛-inch-deep kerf on the diagonal lines on one end. On the opposite end

use a scratch owl to center-punch a ⅛-inch-deep hole where the lines intersect.

4. Remove the live center from the headstock of the lathe, using the rod designed for this purpose.
5. Locate the live center into the precut saw grooves (kerf). To set the prongs of the live center into the grooves firmly tap it with a mallet.
6. With the spur firmly imbedded in the stock remove the live center from the stock and place it back in the lathe headstock.
7. Put a few drops of lubricating oil in the ⅛-inch-deep hole punched in the opposite end of the stock.
8. Position the wood with the grooved end against the live center while holding the stock in position with your free hand.
9. Draw the tailstock to within approximately 1 inch of the stock. Clamp the tailstock to the lathe bed.
10. Revolve the handwheel on the tailstock so that the dead center is driven into the ⅛-inch-deep hole.
11. Secure the stock tightly with the hand wheel on the tailstock, then lock the handwheel adjusting lever.
12. Adjust the tool rest approximately ⅛ inch above the center of the stock. Allow approximately ⅛-inch to ¼-inch clearance between the tool rest and the turning. Revolve the turning to check for proper clearance.
13. Pieces of stock two inches square or more should have the corners removed before inserting the piece in the lathe. Use a plane to take off corners.
14. Select a slow cutting speed, then start the motor.

15. Place a large gouge on the tool rest. Hold the handle toward the end with the right hand. With the left hand hold the blade and guide it along the tool rest. Work from the center towards the ends. Continue rough cutting until the wood is the shape desired.
16. For finish turning increase the speed to 1400 to 1800 r.p.m. The higher speed is used for applying finish and polishing. Use a wide skew chisel for finish cutting.
17. Remove the tool rest for sanding.

18. After a final scraping cut, use medium to fine garnet paper. After a final shearing cut, only a minimal amount of sanding is necessary.
19. Sand all shoulder cuts first with the sandpaper folded. Hold the paper evenly so that the shoulders do not get rounded.
20. Apply finish with a pad of cloth as mentioned under faceplate turning.

**Safety Considerations**

1. In choosing stock for turning avoid loose knots, checks, splintery wood, or improperly glued pieces.
2. Check often to see that the locking adjustments on the tailstock assembly does not become loose.
3. Do not wear loose clothing like neckties, necklaces, or untied cuffs.
4. Wear clear safety goggles.
5. Keep the turning chisels sharp.
6. Rotate stock by hand to check for clearance.
7. Check to see that stock is properly centered.
8. Lubricate the dead center often with lubricating oil.
9. Always hold the tool firmly in both hands.
10. Maintain a well-balanced footing.
11. Rough cutting should be done at slow speed. Advance the speed as the work smooths out.
12. Avoid heavy cuts.
13. Keep hands off rough stock that is revolving.
14. Tools should be kept sharp and stored in a proper location.
15. Tool rest should be removed before sanding or finishing.
16. Always stop the lathe for any measurements.

# 80

# WOODCRAFT

# PROJECTS

## Project 1

# CONTEMPORARY CHEST OF DRAWERS

### Materials Required

*Wood of Your Choice*

**2 sides** ¾″ × 16″ × 42″
**1 top** ¾″ × 15¾″ × 42½″
**4 solid drawer divider dust panels** ¾″ × 15½″ × 42½″
**1 base kickboard** ¾″ × 4″ × 42½″
**1 rail mounted below the top** ¾″ × 2″ × 42½″
**1 back** ¼″ × 41″ × 41¾″
**1 flush front drawer unit** (overall size)
7⅞″ × 15½″ × 42¼″

### Procedures

1. Begin by edge-gluing enough stock to reach the required width for all pieces.
2. Lay out and cut the two sides and top.

   Cut a ¼-inch by ⅜-inch rabbet on the rear inside edges (this will receive the plywood back). When cutting the rabbet, make sure to stop 1 inch from the top of the sides so the rabbet does not show on the extended side pieces. Use an electric router with a rabbet bit or a table saw for this operation.

   Lay out and drill the ⅜-inch counterbored pilot screw holes for the sides to be fastened to the drawer dividers and top. An option is to cut matching dadoes into the sides.
3. Lay out and cut the top and four drawer divider panels to the suggested size.

# Contemporary Chest Of Drawers

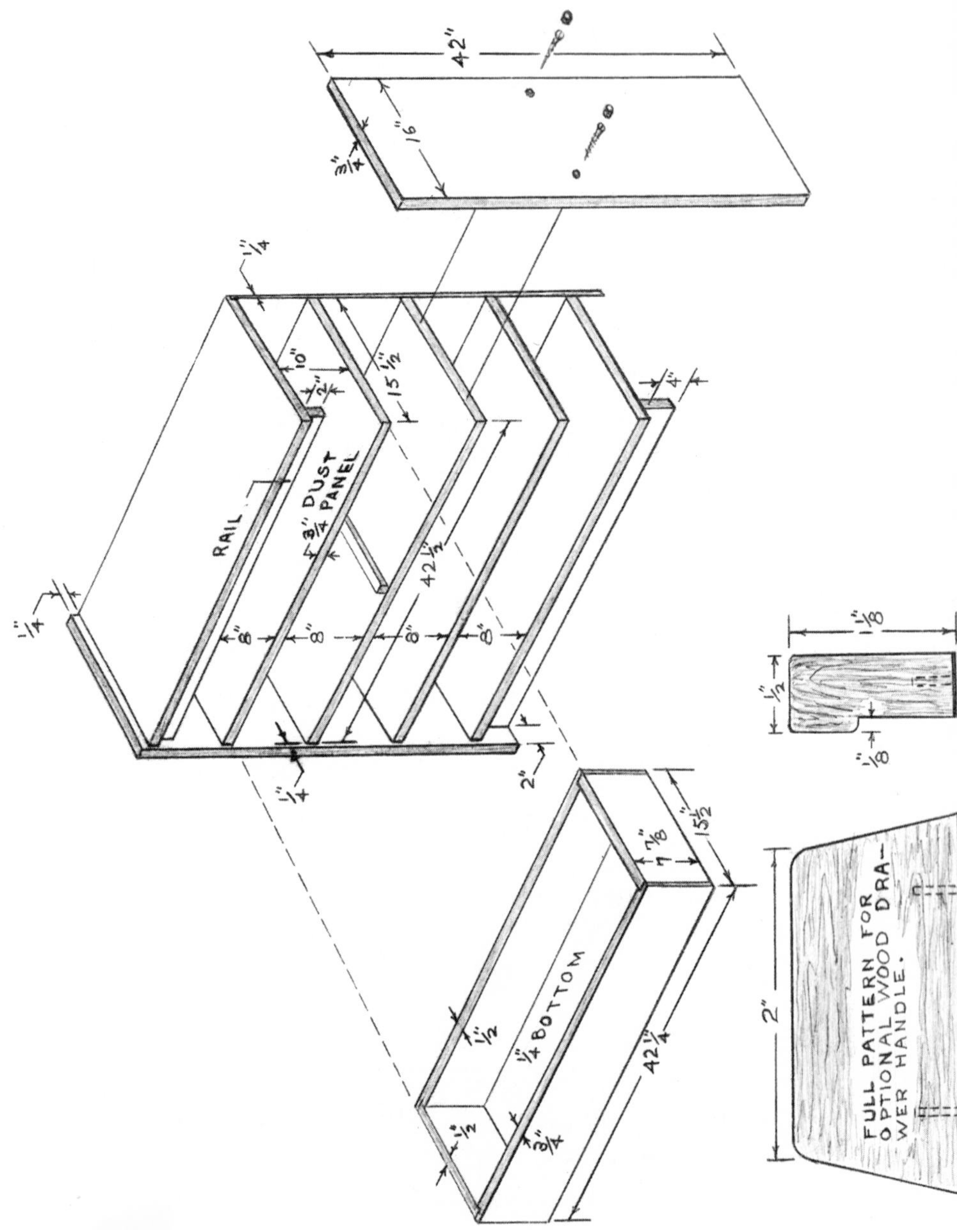

Fasten the sides to the drawer divider panels and top with #10, 1½-inch flathead wood screws. Make sure the sides extend over the top ¼ inch and that the top and drawer dividers are recessed back from the front edge of the side pieces by ¼ inch.

*Note:* Drawer dividers can be made from ¾-inch plywood with a ⅜-inch by ¾-inch strip of pine glued to the front edge.

4. Lay out and cut the back to size, then fasten it to the rear precut rabbets in the top and side pieces.

   Using a carpenter's square, check to see if the assembly is square (90 degrees) before fastening the back permanently with ¾-inch finishing brads.

5. Lay out and cut to size the ¾-inch by 2-inch by 42-inch rail along with the bottom 4-inch base kickboard.

   Recess the rail approximately ¼ inch from the front edge of the top, then glue and clamp it into place.

   Recess the base kickboard 2 inches, then glue and clamp it to the underface of the bottom drawer divider.

6. Construct the drawer units to the overall size indicated in the drawing. Refer to the front of the book for helpful hints on constructing drawers and drawer guides.

7. Scrape off all traces of glue, set and fill all nail holes, plug all counterbored holes, then sand the chest to a smooth surface.

8. Select and apply the colored stain or paint of your choice. When dry, apply several coats of lacquer. Rub the final coat down with paste wax applied to a pad of steel wool.

*Note:* Use at least four coats of spray lacquer. Preserve the finish with an occasional coat of paste wax and buffing.

9. Select and apply the drawer pulls of your choice.

   *Note:* If the pattern in the drawing calls for wood handles, use a back saw to make the ⅛-inch cut for the finger grip. If using a table saw, make the ⅛-inch cut on a ½-inch by 1⅛-inch by 18-inch piece of cherry, then cut eight patterns out with a sabre saw. To attach the handles drill two pilot holes into each handle, then locate and drill two anchor holes through the drawer and fasten with #10, 1¾-inch oval head screws.

## Project 2

# OCTAGONAL MIRROR

### Materials Required

*Wood of Your Choice*

**4 boards** 1″ × 10″ × 18″
**8 dowels** ⅜″ dia. × 3″

### Procedures

1. Lay out and cut four pieces of 1-inch by 10-inch by 18-inch stock. It may be necessary to edge-glue the stock to achieve the 10 inches in width. With a square, check the ends of each piece for a 90-degree cut.
2. With the joint of each board laid out in a staggered pattern, mark the holes in each piece to receive the dowels. The marks should be laid out across the edges and exactly on ½-inch center of the 1-inch-thick edge. Use a marking gauge as a helpful tool to lay out the holes.
3. With ⅜-inch diameter bit, bore each of the sixteen holes to a depth of 1½ inches. Be sure to bore the holes straight, creating a 90-degree angle. Use a doweling jig as a guide.

   Apply two coats of glue to the end grain of the boards, then insert the dowel pins into the prebored holes with glue. Clamp the four pieces with bar clamps until the glue dries.
4. Make the octagonal pattern, then continue to lay it out on the four staggered boards. Cut the frame out with a hand saw or portable sabre saw.

   *Note:* The inner cut out can be round or octagonal in shape.

# Octagonal Mirror

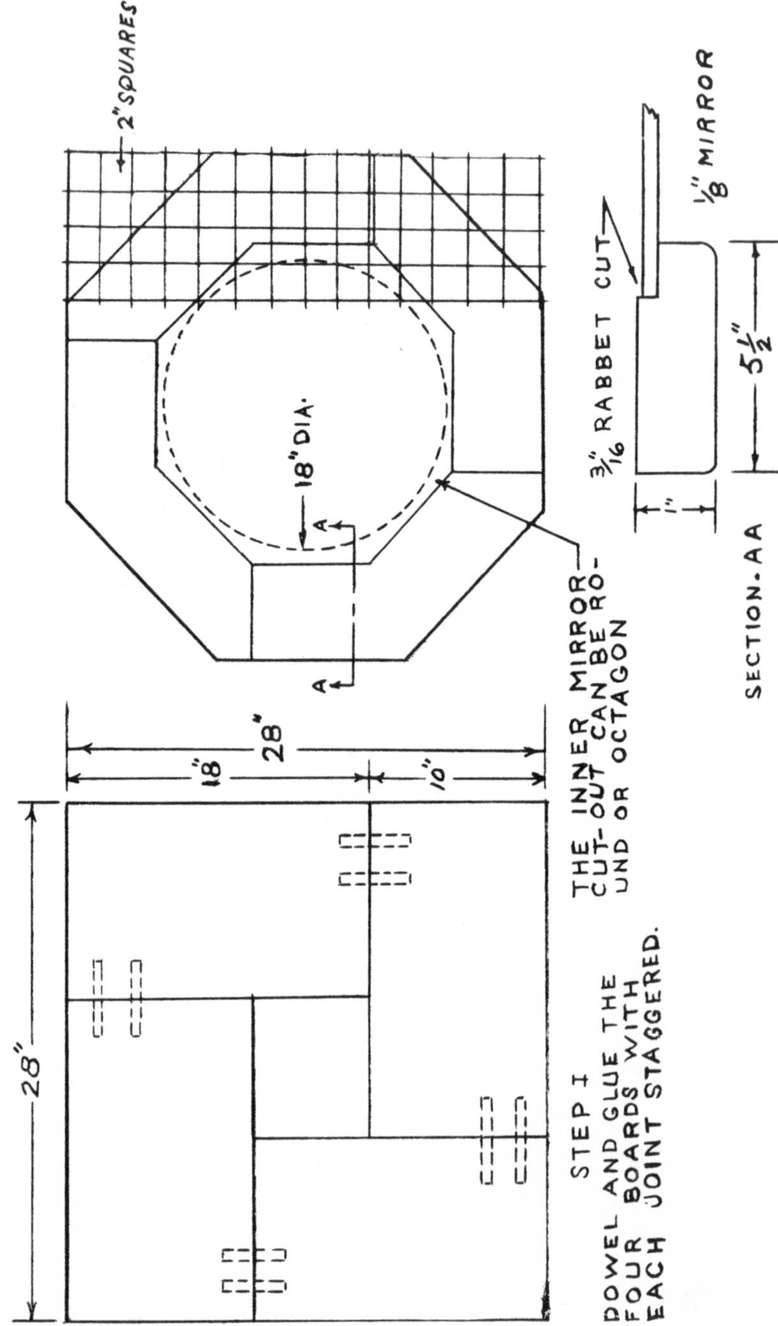

5. With a rabbet bit installed in an electric router make a ¼-inch by ⅜-inch rabbet cut on the back side of the frame. This cut will hold the mirror in place.

   *Note:* If the rabbet cut cannot be made, the mirror can be fastened to the back by taping and tacking it onto the wood.

6. File the rough marks on the inner and outer edge, then sand the project to a smooth surface.

7. Apply several coats of clear polyurethane to the frame, rubbing between coats with 320 emery paper or fine steel wool. Protect the surface with a coat of paste wax and buff.

8. The mirror should be cut and installed into the frame at your local glass supply store.

   *Note:* If you do not wish to use the staggered joint design, simply edge-glue enough stock to reach the 28-inch width, then cut out the pattern.

## Project 3

# ENTRANCE TABLE

### Materials Required

*Wood of Your Choice*

**1 top** ⅝″ × 20″ × 30″
**2 bases to legs** ¾″ × 8″ × 20″
**2 top cleats to legs** ¾″ × 2″ × 18″
**2 middle sections of legs** ¾″ × 6″ × 20″
**1 center cross rail** ¾ × 2″ × 22″
**2 diagonal support braces** ¾″ × 2″ × 9¼″
**1 dowel** ½″ dia. × 16″
**4 drawer runners** ⅜″ × ⅜″ × 7″
**2 smaller drawer runners** ⅜″ × ⅜″ × 5″

*Drawer:*
**1 front** ½″ × 3″ × 23⅛″
**1 back** ½″ × 3″ × 23⅛″
**2 ends** ½″ × 3″ × 8″
**1 bottom** ¼″ × 7½″ × 22⅝″
**1 knob** 1″ dia.
**1 drawer pull (optional)** ½″ × ¾″ × 1″

### Procedures

1. Edge-glue the ¾-inch and the ⅝-inch stock together to make necessary widths for all parts.
2. Lay out and cut the three pieces of stock for each leg assembly to the size and design shown in the drawing. Approximately 1 inch from the bottom of the middle post of each leg bore a ½-inch-diameter hole with a spade bit to a depth of 5 inches, then insert with glue a ½-inch-diameter dowel.

*Note:* The dowel is inserted so that when screwing into the end grain, the screws will attach to the dowel for more holding strength.

3. Drill three ½-inch counterbored pilot screw holes through the bottom of the base and into the middle post, then secure the base to the post with #12, 3-inch flathead wood screws.

   Drill two ⅜-inch counterbored pilot screw holes through the top cleat and into the post, then fasten the cleat flush to the top outer face of the post with #12, 1½-inch flathead wood screws.

4. Cut the stock for the cross rail to the size suggested.

   At approximately 1 inch from each end of the cross rail bore a ½-inch-diameter hole completely through, then insert with glue a ½-inch-diameter dowel.

   Locate the cross rail to the center of the two legs, then drill two ⅜-inch-diameter counterbored pilot screw holes through the cleat and post and into the cross rail. Secure the two legs to the cross rail with #12, 3½-inch flathead wood screws.

5. Lay out and cut the diagonal supports to the exact dimensions shown in the drawing.

   Fasten the support braces to the cross rail and leg by drilling countersunk pilot holes through each angled end then by securing them to the members with #12, 1½-inch flathead wood screws.

6. Lay out and cut the top to the recommended size and shape.
   The top edge should be rounded over approximately ⅛ inch. This can be achieved by planing, filing, then sanding until it is round.

# Entrance Table

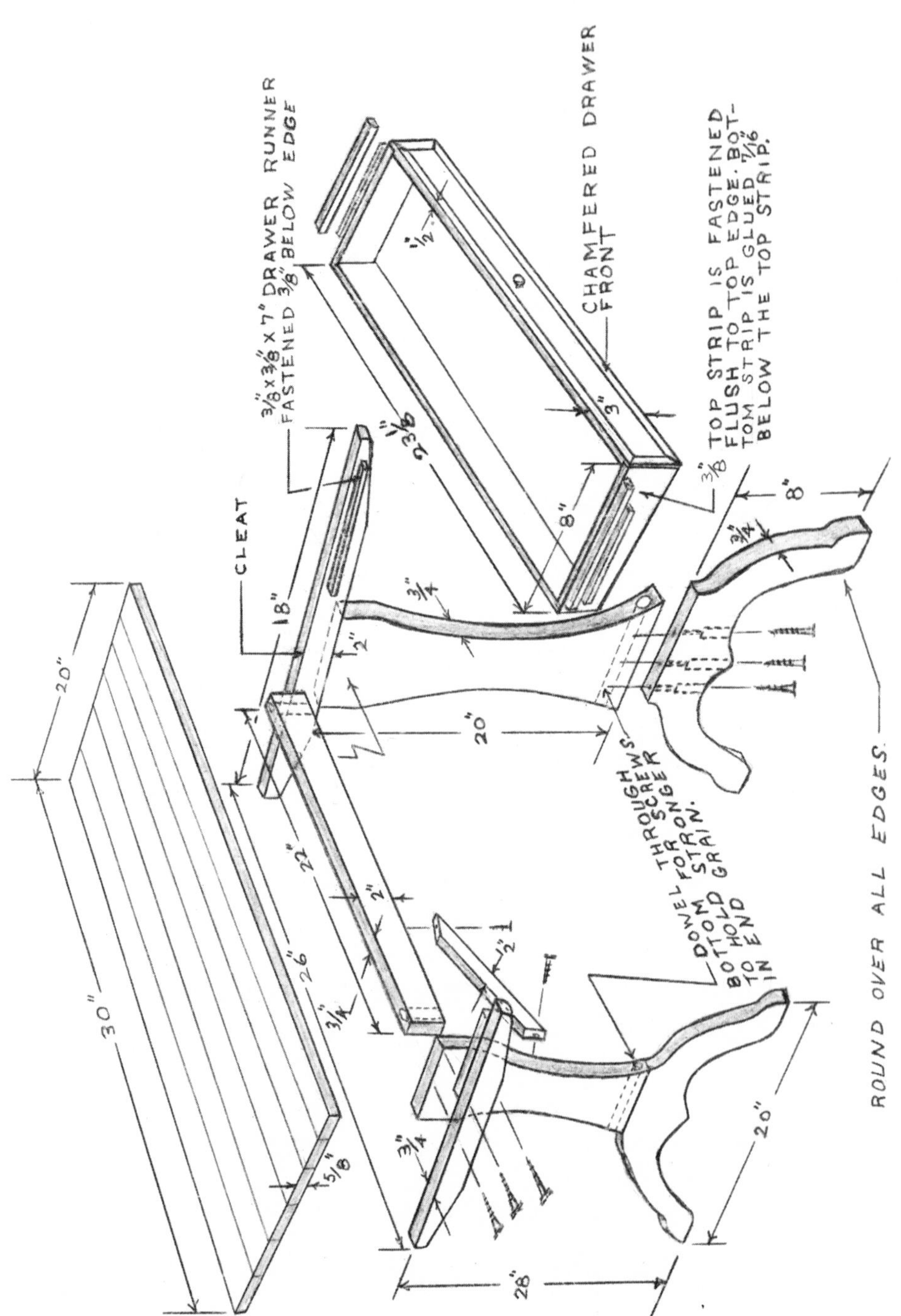

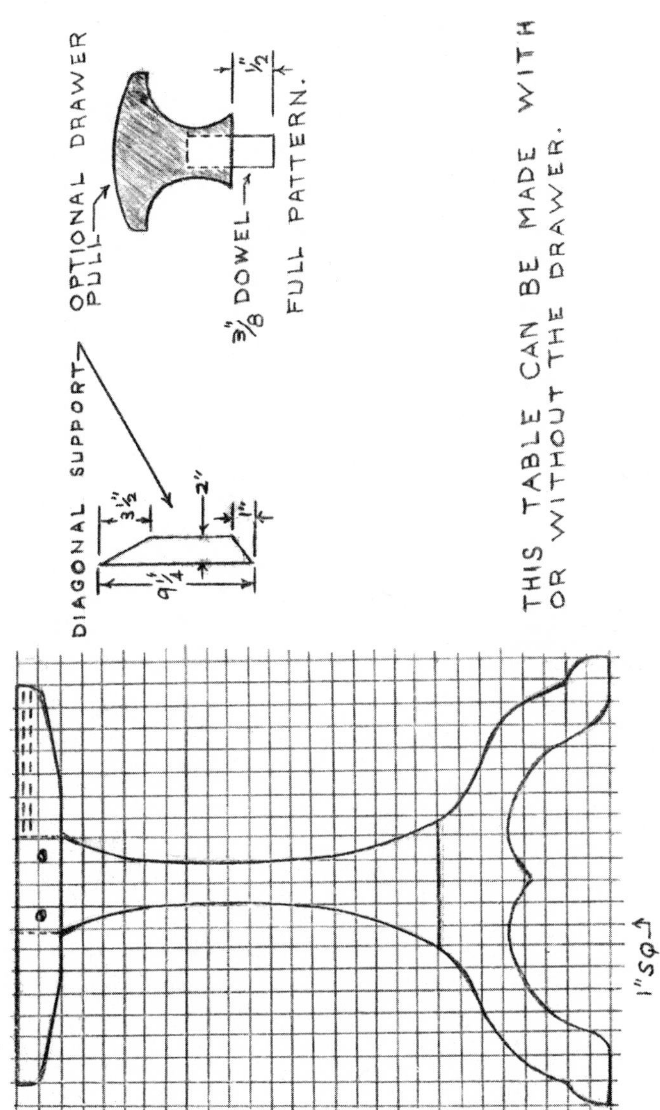
DIAGONAL SUPPORT
9 1/4"
3 1/2"
2"
1"
OPTIONAL DRAWER PULL
3/8" DOWEL
1/2"
FULL PATTERN.
THIS TABLE CAN BE MADE WITH OR WITHOUT THE DRAWER.
1" SQ.

Attach the top to the leg assembly by drilling four countersunk pilot holes up through the cleats and into the underside of the top piece. Fasten with #12, 2-inch flathead wood screws.

7. For the drawer construction lay out and cut all pieces to the suggested size.

   Lay out and cut two ¼-inch by ½-inch rabbet joints on both front and back pieces.

   Lay out and cut a ¼-inch by ¾-inch chamfer on the front piece of the drawer. This can be done with a hand plane or a table saw.

   Cut a ¼-inch groove approximately ½-inch from the bottom edge on all four pieces to receive the plywood bottom.

   Fasten all the pieces together to complete the drawer assembly with glue and 1-inch finishing brads.

8. Glue a ⅜-inch by ⅜-inch by 7-inch rider strip flush to the top edge of each end of the drawer.

   Glue a ⅜-inch by ⅜-inch by 5-inch guide strip approximately 7/16 inch below the top strip.

   Glue a ⅜-inch by ⅜-inch by 7-inch drawer runner strip approximately 7/16 inch below the top edge of each leg cleat.

   *Note:* After the finish is applied to the runners and strips, apply soap to them so that the drawer will move smoothly.

9. Plug all exposed counterbored holes with ⅜-inch-diameter furniture buttons or flush dowel plugs. Thoroughly sand the entire table to a smooth surface, then select and apply the stain of your choice. When dry, apply a minimum of four coats of finishing oil, rubbing lightly between dry coats with fine steel wool or pumice and oil. Protect the final finish with a coat of paste wax and buff.

## Project 4

# PORTABLE SERVING STAND

### Materials Required

*Wood of Your Choice*

*"X" Stand:*
**4 legs** ¾″ × 2″ × 37″
**2 bottom rungs, dowels** ¾″ dia. × 16½″
**3 middle and top rungs, dowels** ¾″ dia. × 18″
**1 yard of webbing for straps**

*Note:* The webbing straps can be purchased at your local fabric store or department store.

*Tray:*
**1 bottom** ¾″ × 20″ × 26″
**2 sides handles** ¾″ × 2¾″ × 18″
**1 back** ¾″ × 2¾″ × 23″

### Procedures

1. Make patterns then proceed to lay out and cut all of the parts to the design suggested. A sabre saw or hand saw can be used to accomplish the cutting of scrolled parts.
2. Assemble the X-shaped legs first.

   Begin by using a ¾-inch spade bit to bore a ¾-inch-diameter hole completely through the center of the two inner legs.

   Continue to bore ¾-inch-diameter holes to a depth of ½ inch for all other holes. Be sure to align the holes exactly.
3. Fasten the three ¾-inch-diameter dowels into the prebored ¾-inch holes for the two inner legs.

# Portable Serving Stand

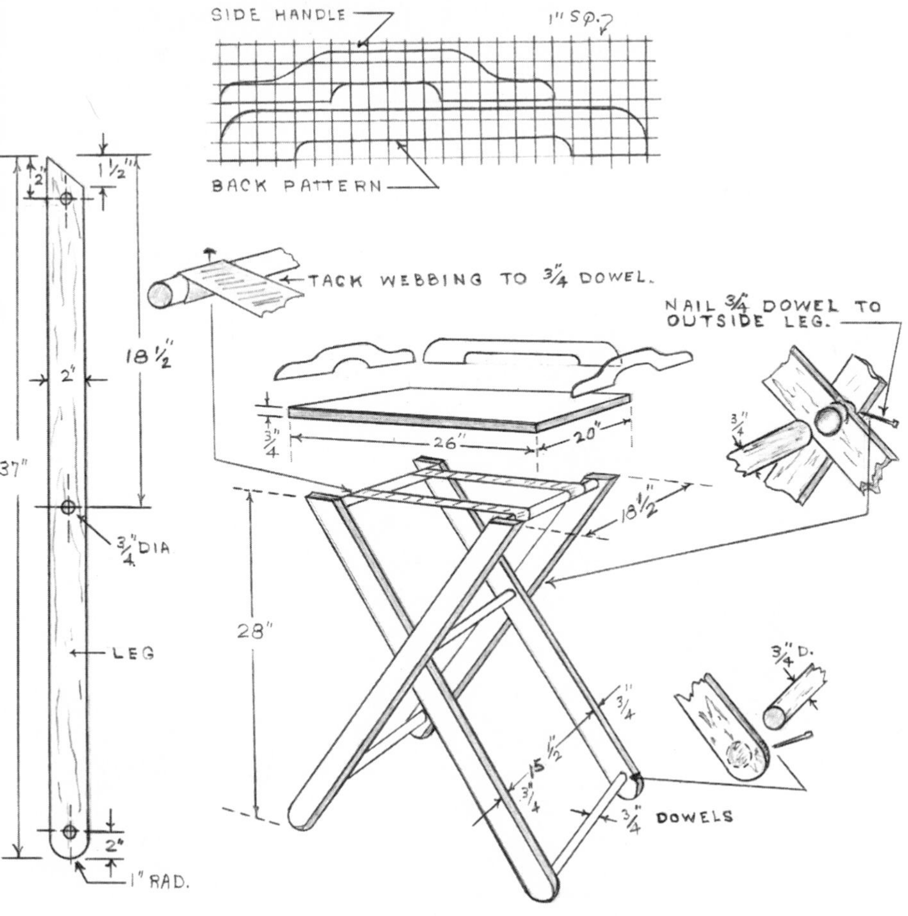

Complete the leg assembly by fastening the outer two legs together with ¾-inch dowels on the top and bottom rung. Then place them over the pivotal middle dowel placed in the center of the two inner legs. The center pivotal ¾-inch-diameter dowel should extend out ½ inch from each side of the inner legs to receive the two outer legs.

4. The center pivotal dowel is held stationary to the outer legs by driving a #4 finishing nail through the edge of the leg and into the ¾-inch dowel (do not nail the middle dowel to the inner legs). Hold all other dowels stationary by driving a #4 finishing nail through the sides and into the dowels.

   *Note:* An option to using a nail to hold the dowel rungs in position is to drill a ¼-inch-diameter hole through the legs and dowel rung and then insert a ¼-inch-diameter dowel pin.

5. On one side of the X-shaped pivotal legs attach two 3-inch by 26-inch straps of leather or webbing by wrapping them around the dowel twice and then tacking them in place.

   Proceed to open the X-shaped legs to approximately 22 inches, then fasten the opposite ends of the straps following the procedure described above.

   *Note:* The straps should be fastened on after the finish has been applied to the stand.

6. Assemble the precut pieces for the serving tray by gluing and clamping. For additional strength drill two countersunk pilot screw holes up through the bottom and into each piece, then screw in #12, 1½-inch flathead wood screws.

7. Set and fill all nail holes, scrape off all traces of glue, then proceed to sand the serving stand to a smooth surface. Do the final sanding of the top by hand so that all machine marks are eliminated.

8. Select and apply the colored stain of your choice. When thoroughly dry, apply at least four coats of clear finishing oil, making sure to rub lightly between dry coats with 320 grit emery paper or fine steel wool.

## Project 5

# CHARLOTTE'S BENCH

### Materials Required

*Wood of Your Choice*

**1 top** ¾″ × 9½″ × 38″
**2 legs** ¾″ × 9½″ × 15¾″
**2 side rails** ¾″ × 5″ × 32″

### Procedures

1. Glue the stock to obtain the required width for the top and legs.

2. Lay out the patterns, then cut the five pieces to the size and design recommended.

3. Lay out and cut the four ¾-inch by 1½-inch notches into each leg. The notches can be cut out by making a number of release cuts to a ¾-inch depth with a hand saw and then using a chisel and mallet to clean out the cut.

4. Lay out and cut the two ¼-inch by ¾-inch dado cuts into the top piece. Then proceed to fasten the two legs into the dado joints using glue and #6 finishing nails.

5. Finish the assembly by attaching the two side rails to the precut ¾-inch by 1½-inch notches. Use glue and bar clamps for fastening. If added reinforcement is desired, insert dowel pins through the rails and into the legs.

6. Set and fill all holes, then sand the bench to a final smooth surface.

# Charlotte's Bench

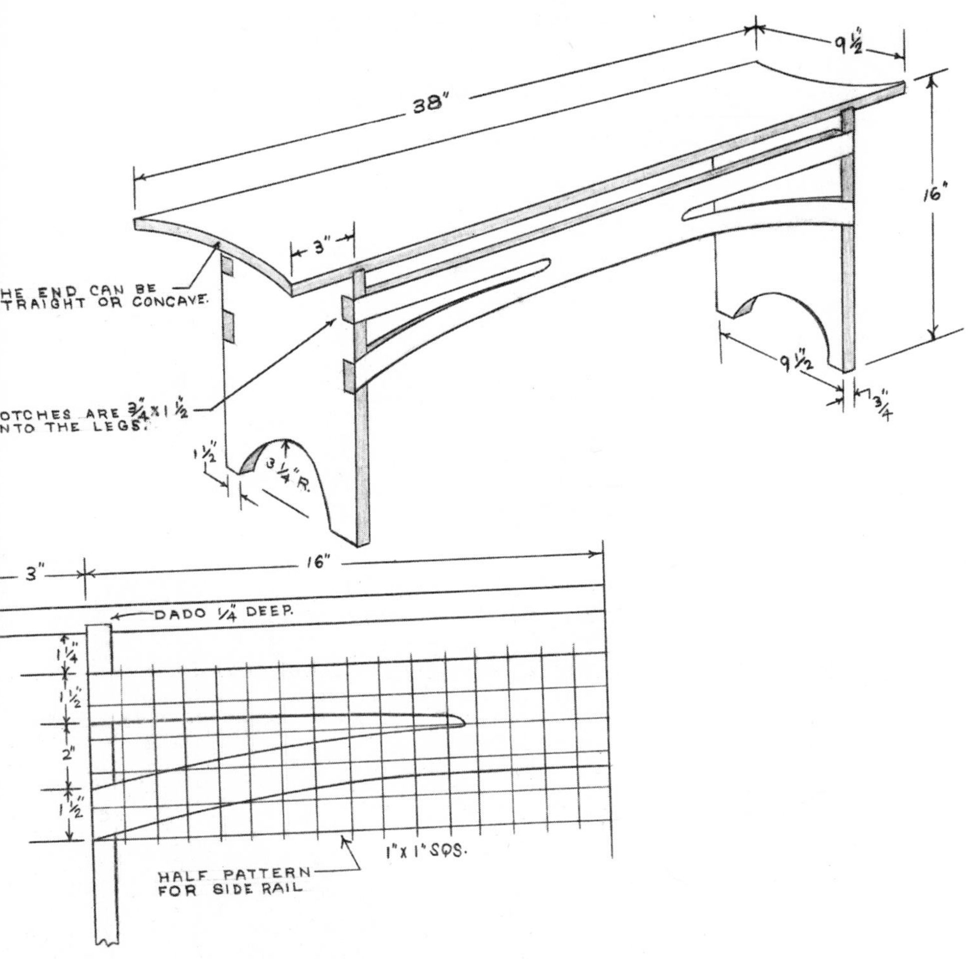

7. Select and apply the colored stain or paint of your choice. When dry, apply several coats of clear lacquer. Finish rubbing lightly between coats with fine steel wool or 320 emery paper. Preserve the finish with an occasional coat of paste wax or lemon oil.

## Project 6

# CANDLE STAND (PLANT)

### Materials Required

*Wood of Your Choice*

**1 top** ¾″ × 13″ × 13″
**1 cleat** ¾″ × 2″ × 10″
**1 center post** 2¼″ × 2¼″ × 22″
**2 bases** 2″ × 2″ × 12″
**1 dowel pin** ½″ × 1¼″
**1 dowel pin** ½″ × 1¾″

### Procedures

1. Edge-glue and clamp the stock to the size indicated. Lay out and cut the top piece to the suggested design.
2. Cut a 45-degree bevel on both ends of the 2-inch by 10-inch cleat, then fasten the cleat beneath the top by gluing and clamping.
3. Lay out and cut the taper on the center post. After planing the four sides even, lay out and cut the ⅜-inch chamfer on the four edges of the center post.
4. Before cutting the design on the two base pieces, lay out and cut the middle half-lap joint, then lay out and cut the design on the two base pieces. Fasten the two base pieces to the middle half-lap joint by gluing and clamping.
5. Bore a ½-inch hole to a depth of 1¼ inches in through the cleat and top. Bore a ½-inch hole to a depth of 1¾ inches through the center of the base.

# Candle Stand (Plant)

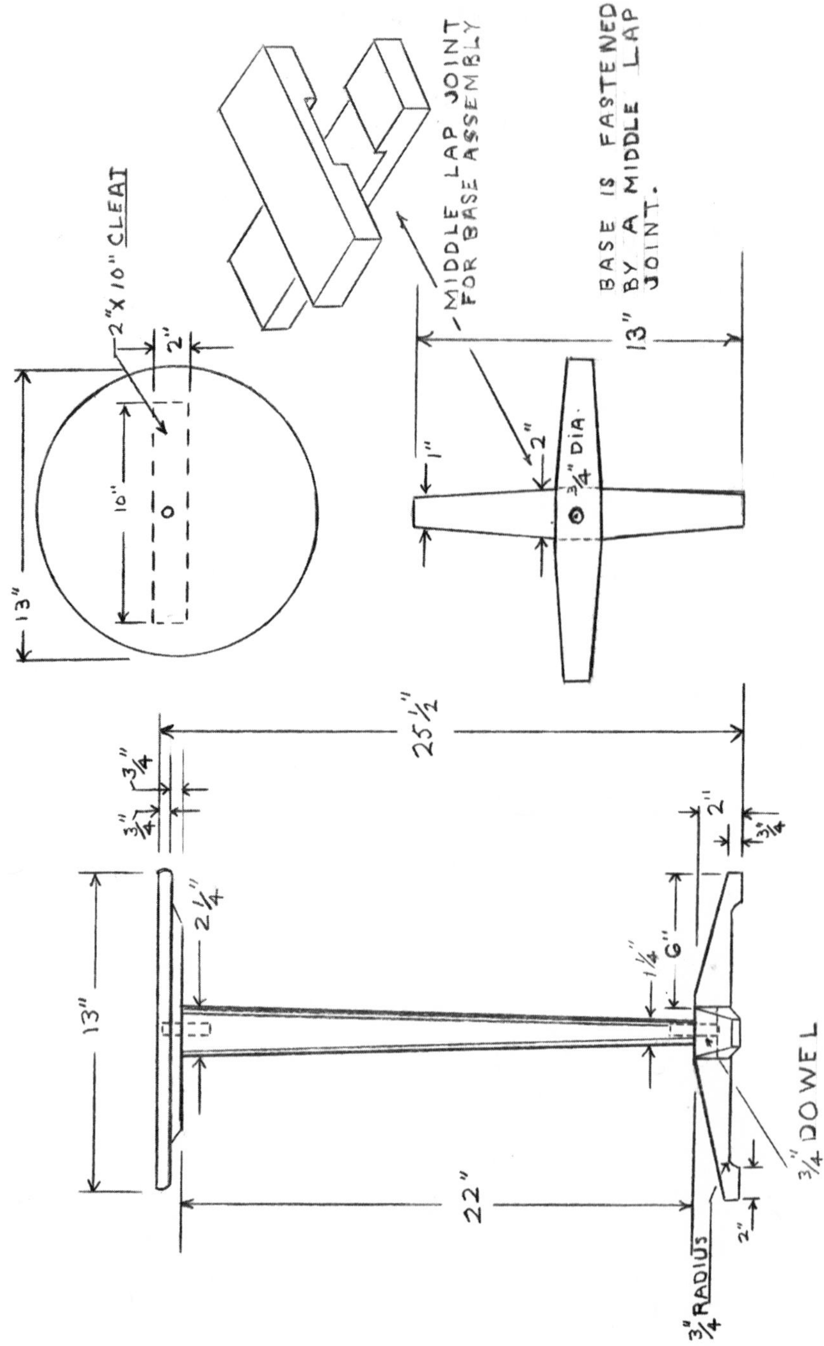

6. Assemble all of the pieces with ½-inch-diameter dowel pins glued into the prebored holes.
7. Scrape off all traces of glue, then sand the stand to a smooth surface.
8. Select and apply the stain or paint of your choice. When dry, apply several coats of finishing oil, rubbing between coats with fine steel wool or pumice and oil. Protect the finish with a coat of paste wax; buff.

## Project 7

# MEDICINE CHEST

### Materials Required

*Wood of Your Choice*

**2 top and bottom pieces** ¾″ × 5¾″ × 25¾″
**2 shelves** ¾″ × 4¼″ × 22¾″
**1 inner top frame** ¾″ × 4¼″ × 22¾″
**1 inner bottom frame** ¾″ × 4¼″ × 22¾″
**2 stiles** ¾″ × 2″ × 25½″
**1 back (optional)** ¼″ 23¼″ × 26″

*Door:*
**2 top and bottom rails** ¾″ × 2″ × 20¾″
**2 stiles** ¾″ × 2″ × 25⅜″
**2 mirrors** ½″ × 16″ × 20½″ Available at your local glass dealer.
**1 knob** 1¼″ dia.
**H-surface hinges** 3″

### Procedures

1. Lay out and cut all stock to the size shown in the drawing.
2. With a table saw or a router with a rabbet bit cut a ¼-inch by ⅜-inch rabbet on the back inner edge of the four sides of the cabinet. This rabbet cut will receive the ¼-inch back.
3. Lay out to position, then fasten the two inner shelves along with the inner top and bottom frame to the two sides of the cabinet. Use #6 finishing nails.

   Continue to attach the ¼-inch back into the precut rabbet with 1-inch finishing nails.

   *Note:* Be sure the cabinet is square before fastening the back.

# Medicine Chest

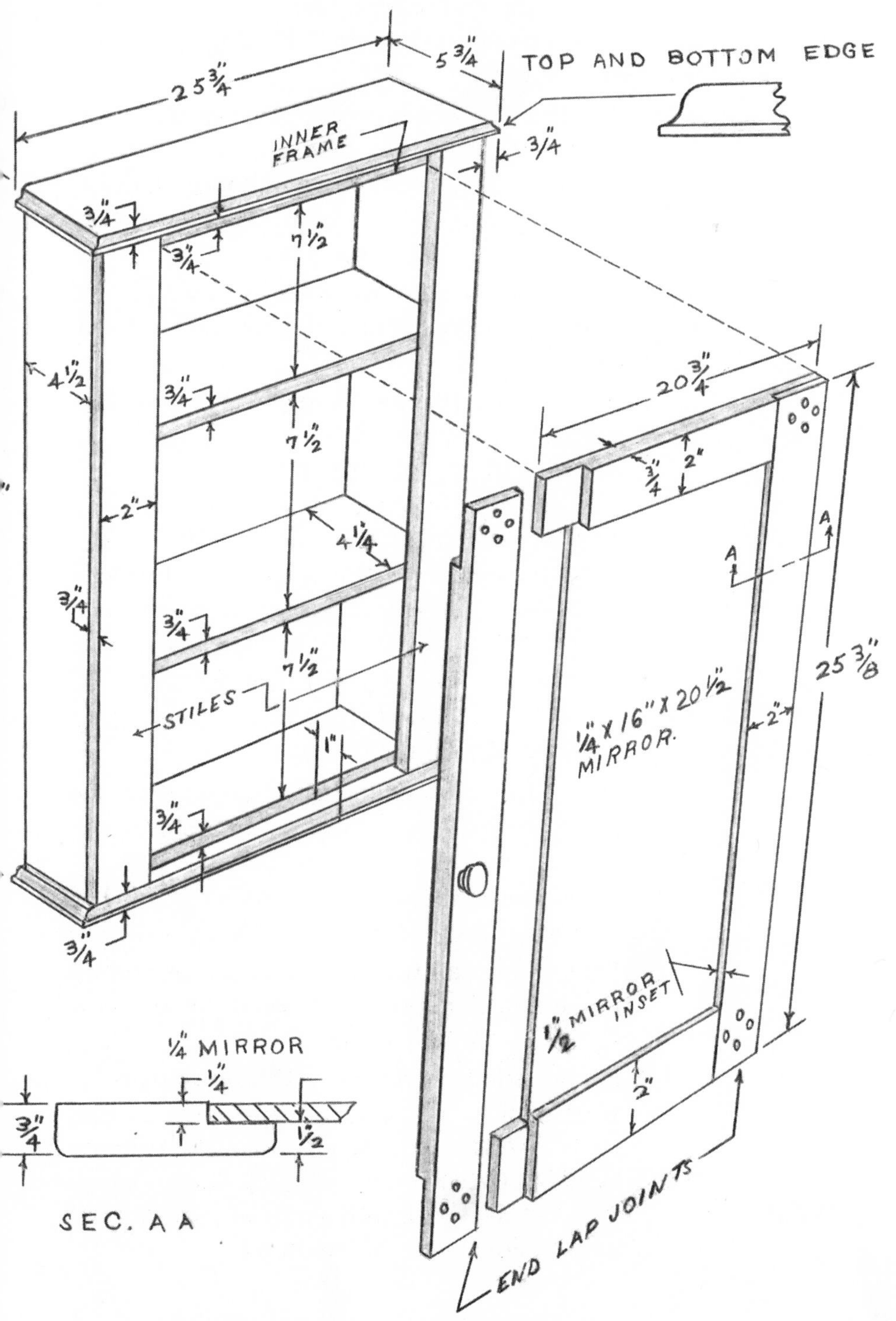

4. With a router and roman ogee bit cut the trim on the top and bottom pieces. *Note:* The edge trim is equally attractive if it is simply rounded over slightly.

   Continue to attach the top and bottom to the face of the frames by using glue and #6 finishing nails. Be sure the top and bottom has a 1-inch overhang on the front of the cabinet and a ¾-inch overhang on the sides.
5. Finish the cabinet assembly by fastening the two stiles to the front with glue and #6 finishing nails.
6. To construct the door begin by laying out and cutting a ¼-inch by ⅜-inch rabbet on the inner edge of the four pieces.

   Proceed to lay out and cut a ⅜-inch by 2-inch half-lap joint at the ends of each of the four door pieces. *Note:* When the door frame is glued and clamped, check it with a square to make sure the corners are at 90-degree angles.
7. Insert the ¼-inch mirror with glass points, or screw ¼-inch wood strips to the inner frame and overlapping the mirror.

   Install the knob and surface hinges onto the frame, then fasten the hinges to the stile of the chest.

   *Note:* Allow 1/16-inch clearance on the top and bottom of the door.
8. Scrape clean all traces of glue, set and fill all nail holes, then proceed to sand the medicine chest to a smooth surface.
9. Select and apply the colored stain or paint of your choice. When fully dry, apply several coats of clear polyurethane finish, rubbing lightly between coats with #320 emery paper. Preserve the finish with a coat of paste wax or an occasional coat of lemon oil.

## Project 8

# CHOPPING BLOCK

### Materials Required

*Wood of Your Choice*

**1 top** 1½″ or 2″ × 24″ × 32″
**4 legs** 2″ × 2″ × 34″
**4 side rails** ¾″ × 2½″ × 18″
**2 back rails** ¾″ × 2½″ × 26″
**2 drawer runners** ⅜″ × ½″ × 18″
**2 runners attached to legs** ⅜″ × ½″ × 20″
**16 flathead wood screws** #12, 3″

*Drawer:*
**1 front** ¾″ × 4″ × 25⅛″
**1 back** ¾″ × 4″ ×25⅛″
**2 sides** ¾″ × 4″ × 19¼″
**1 bottom** ¼″ × 19¼″ × 24⅜″
**2 drawer handles** ½″ × 1″ × 2½″
**2 dowels for handles** ¼″ × 1½″

### Procedures

1. Glue up enough stock to reach the required width and thickness for all pieces necessary.
2. Lay out and cut the top to size. Round over the edges approximately ⅛ inch.
   Lay out and drill the four ⅜-inch-diameter counterbored pilot holes for the screws to be fastened to the top of the legs.
3. Cut the 2-inch by 2-inch by 34-inch legs, then locate and drill the ⅜-inch-diameter counterbored pilot holes for the screws to be fastened to the rails.

   *Note:* Dowel or mortise and tenon joints may be used for fastening the rails to the legs.

# Chopping Block

3/8" x 1/2" DRAWER RIDER. THE DRAWER RIDER IS EVEN WITH TOP EDGE. THE RUNNERS ON THE LEGS ARE 9/16" BELOW THE TOP.

RAIL

TOP VIEW OF LEG AND RAIL CONST. 3" SCREWS PLUGED OVER.

FULL PATTERN FOR DRAWER HANDLE MADE FROM 1/2" STOCK.

1/4" DOWEL GLUED INTO HANDLE THEN INTO DRAWER.

4. Lay out and cut the six rails to the correct size, then construct the leg assembly by fastening the rails to the legs with #12, 3-inch flathead wood screws fastened through the predrilled holes.

5. Lay out and cut the two ⅜-inch by ½-inch by 20-inch drawer guide runners. Secure them 9/16 inch below the top edge of the legs. Use glue and clamps, then reinforce them with a 1-inch finishing brad.

6. Lay out and cut all of the stock for the drawer unit.

   Construct the unit by making ¼-inch by ½-inch rabbet joints on both ends of the front and back pieces.

   Cut a ¼-inch groove approximately ½-inch from the bottom inner face of the four pieces. Insert the ¼-inch bottom into the groove, then fasten the sides into the rabbet joints on the front and back pieces, using glue and #4 finishing nails.

   *Note:* If a simpler method is preferred, fasten the drawer together by butting the corners together and fitting the bottom to the inner sides with nails. Glue the two ⅜-inch by ½-inch by 18-inch riders so that they are flush to the top edge of the drawer.Refer to the book's introductory section on drawer construction.

7. Lay out the handles to the full pattern size given in the drawing, then drill a ¼-inch-diameter hole in the center of the handle and a ¼-inch hole into the place on the drawer where you desire to locate the handle. Insert a ¼-inch-diameter by 1½-inch dowel into the handle and drawer, with glue.

8. Plug all ⅜-inch counterbored holes with ⅜-inch-diameter furniture buttons or ⅜-inch dowel plugs. Set and fill all holes, then sand the project to a smooth surface.
9. Apply at least four coats of oil finish, rubbing between dry coats with fine steel wool. Preserve the finish with an occasional coat of oil finish.

   *Note:* Use soap or paste wax on the drawer runners to reduce friction.

## Project 9

# CUTTING BLOCK

### Materials Required

**7 strips** ½″ × 1⅜″ × 15″ maple
**8 strips** ¾″ × 1⅜″ × 15″ walnut

### Procedures

1. Lay out and cut the fifteen strips to the size suggested.

## Cutting Block

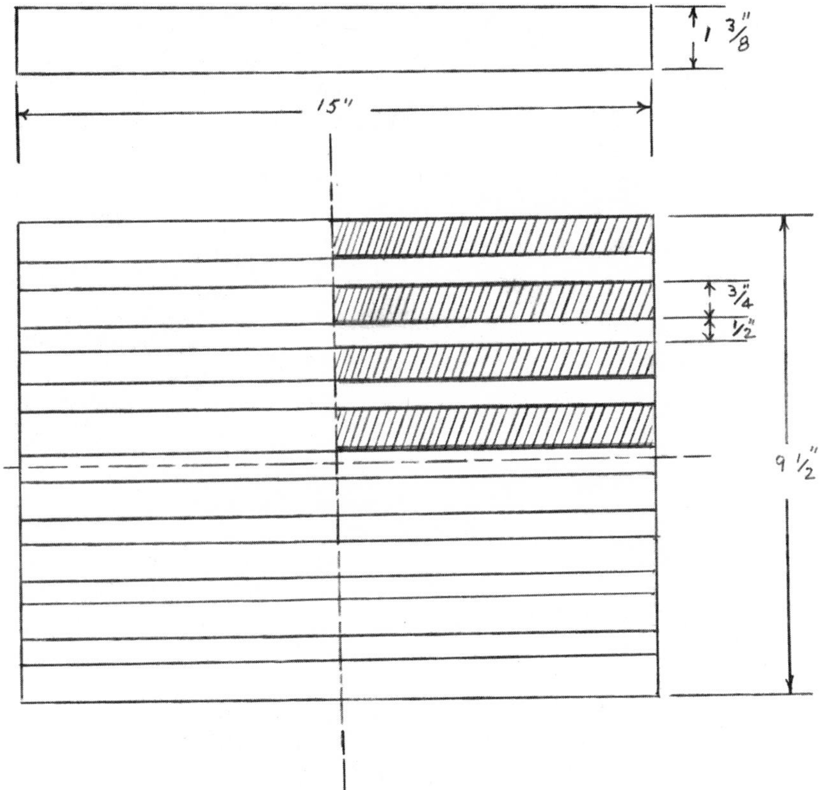

2. Plane the edges to be laminated.
3. Glue and clamp the fifteen alternate hardwood strips.
4. Plane and sand both surfaces to a smooth finish.
5. Apply several coats of clear oil finish, rubbing between coats with fine steel wool or pumice and oil.

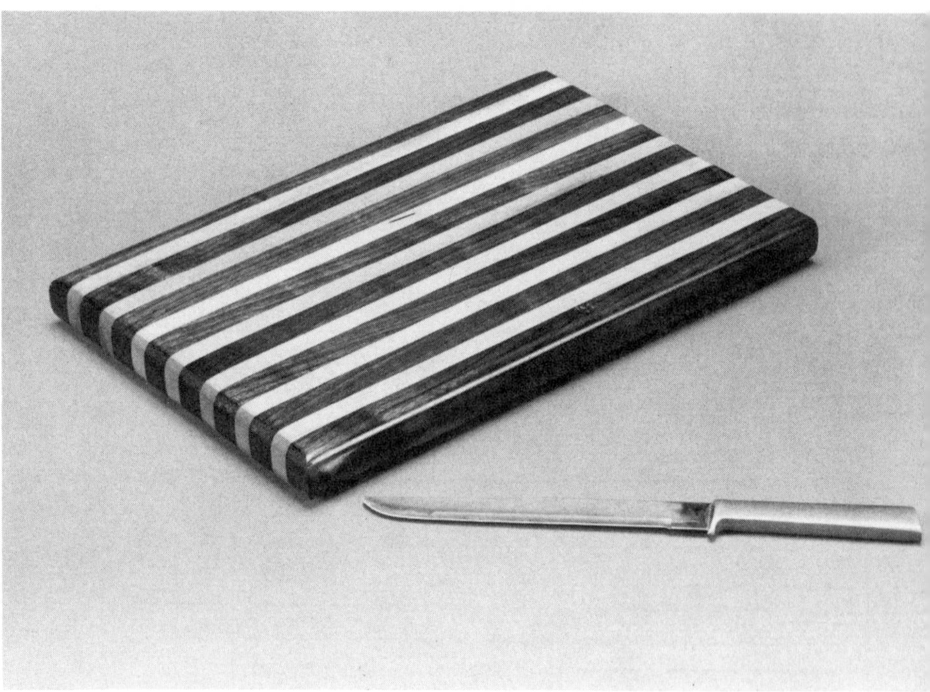

## Project 10

# OUTDOOR WINDOW FLOWER BOXES

### Materials Required

*Wood of Your Choice*

*Material for One Box:*
**1 front** ¾″ × 8″ × 34″
**1 back** ¾″ × 7¼″ × 32½″
**1 bottom** ¾″ × 6½″ × 32½″
**2 end pieces** ¾″ × 7¼″ × 8″

*Note:* If the scalloped box is to be made, reduce the length of each piece above by 2 inches.

### Procedures

1. Select the design, then proceed to make the pattern by enlarging it to 1-inch by 1-inch squares.
2. Lay out the design on the correct size stock, then cut to shape with a sabre saw or coping saw.
3. Lay out and cut the remaining stock to complete the box.
4. Assemble the front of the box to the front edge of both side pieces, then fasten the back to the inner face of the side pieces.

   Secure the bottom snugly within the four sides of the box. As an alternative design, recess approximately ½ inch from the bottom edge of the four sides.

   The box can be assembled with the use of gal-

# Outdoor Window Flower Boxes

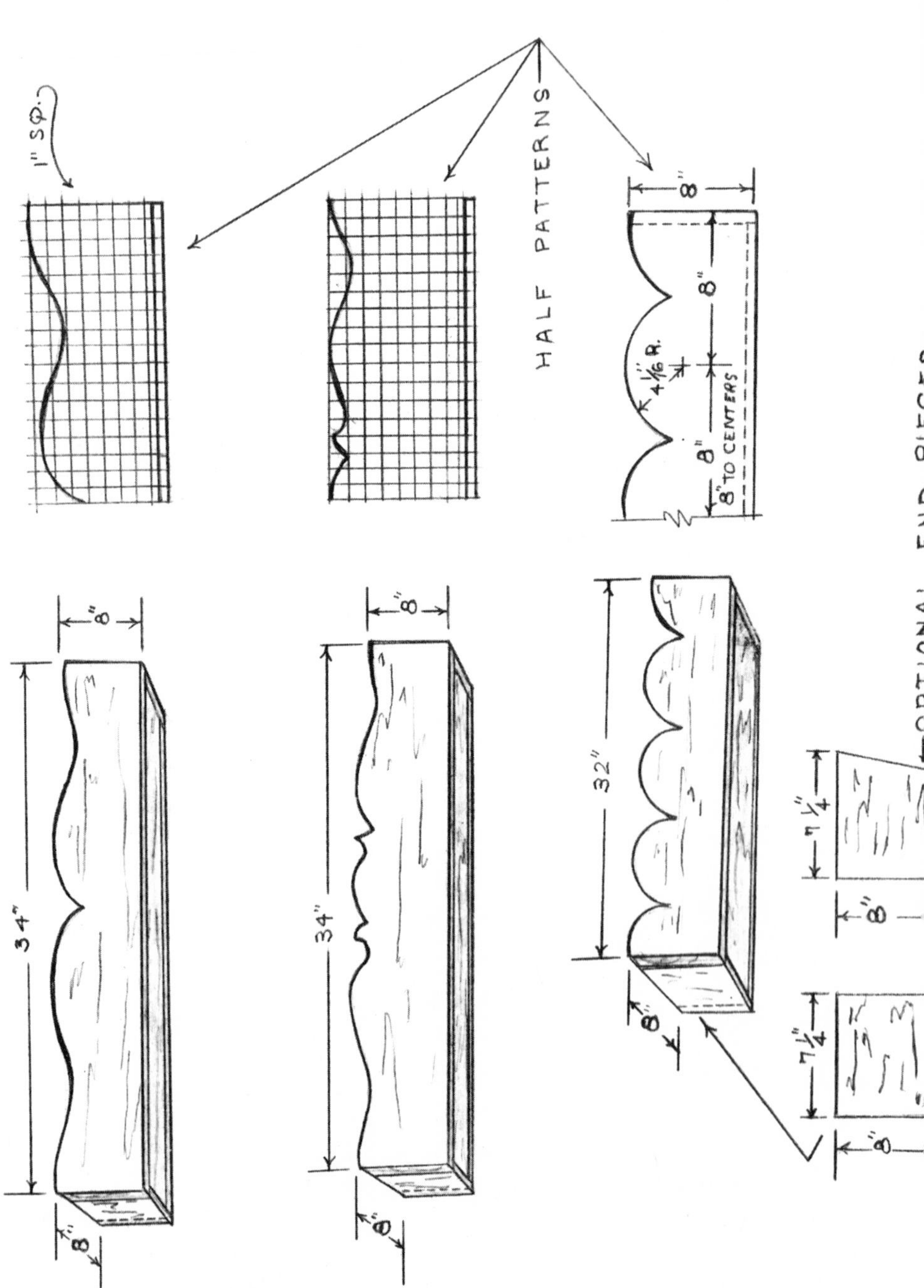

vanized finishing nails or galvanized screws recessed to receive a dowel plug.

*Note:* Galvanized hardware will not rust.

5. Drill three 3/16-inch-diameter holes through the back at approximately 9 inches apart to hang the box.
6. Set and fill all holes, then file and sand the project to a smooth surface.
7. Select and apply the colored outdoor paint of your choice, making sure to use at least two coats for adequate protection. Stain may be applied, but it is important to use at least two coats of spar varnish as a sealer. *Note:* Spar varnish protects against the moisture and salt in the air.

Project 11

# HANGING PLANTER

## Materials Required

*Wood of Your Choice*

**4 sides** ½″ × 5½″ × 5½″
**1 top** ½″ × 5″ × 5″
**1 bottom** ½″ × 5″ × 5″
**finishing nails** #4
**1 eye hook** 1″

## Procedures

1. Lay out and cut all pieces to the size indicated in the drawing. Set the saw to a 45-degree angle, then proceed to make the 45-degree angle cut on the four side pieces for the corner joints.

2. Make a pattern for the design, then transfer it to the four side pieces. Drill a hole into the design layout, then insert the blade of a sabre saw or jig saw to cut out the design.

3. Assemble the four sides together with glue and #4 finishing nails.

4. Fasten the top and bottom with glue and #4 finishing nails. Both pieces should be recessed ¼ inch.

5. Locate the center of the top, then screw in an eye hook for hanging the planter.

6. Clean off all traces of glue. Set and fill all nail holes, then sand the project to a smooth surface.

# Hanging Planter

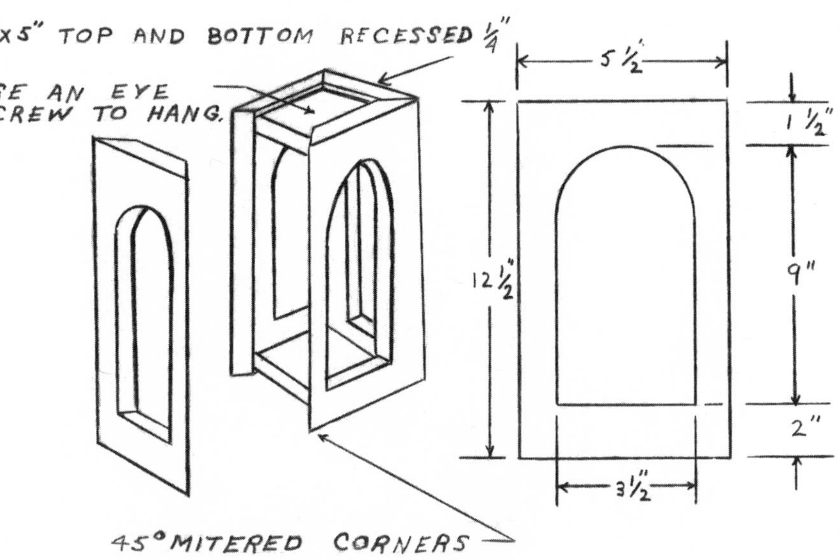


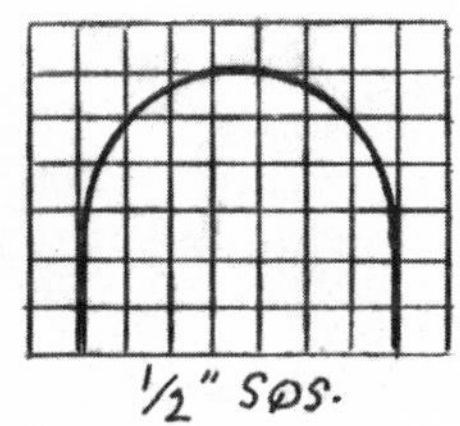

7. Select and apply the stain or paint of your choice. When dry, apply several coats of clear lacquer. Rub between each coat with fine steel wool or pumice and oil. Protect the finish by applying a coat of paste wax and buffing.

Project 12

# HOPE CHEST

## Materials Required

*Wood of Your Choice*

**2 ends** ¾″ × 16″ × 18″
**1 front** ¾″ × 18″ × 46″
**1 back** ¾″ × 18″ × 46″
**1 bottom** ⅜″ × 16″ × 45¼″
**1 top** ¾″ × 18½″ × 47¾″

*Shelf Parts:*
**1 front** ¾″ × 4″ × 44¼″
**1 back** ¾″ × 4″ × 44¼″
**2 ends** ¾″ × 4″ × 6½″
**2 dividers** ¾″ × 3″ × 6½″
**1 bottom** ⅜″ × 6½″ × 43½″

*Base Molding:*
**1 front** ¾″ × 4″ × 47½″
**2 ends** ¾″ × 4″ × 19″

*Note:* An alternative to making the chamfered base molding is to purchase 4″ clamshell molding from your local lumber and mill store.

## Procedures

1. Glue up enough stock to reach the required width for all pieces, then lay out and cut all pieces to the sizes indicated in the drawing.
2. Cut a ⅜-inch by ⅜-inch dado approximately 1 inch above the bottom edge of the end pieces.
3. With glue, insert the ⅜-inch bottom into the precut dadoes in the end.

# Hope Chest

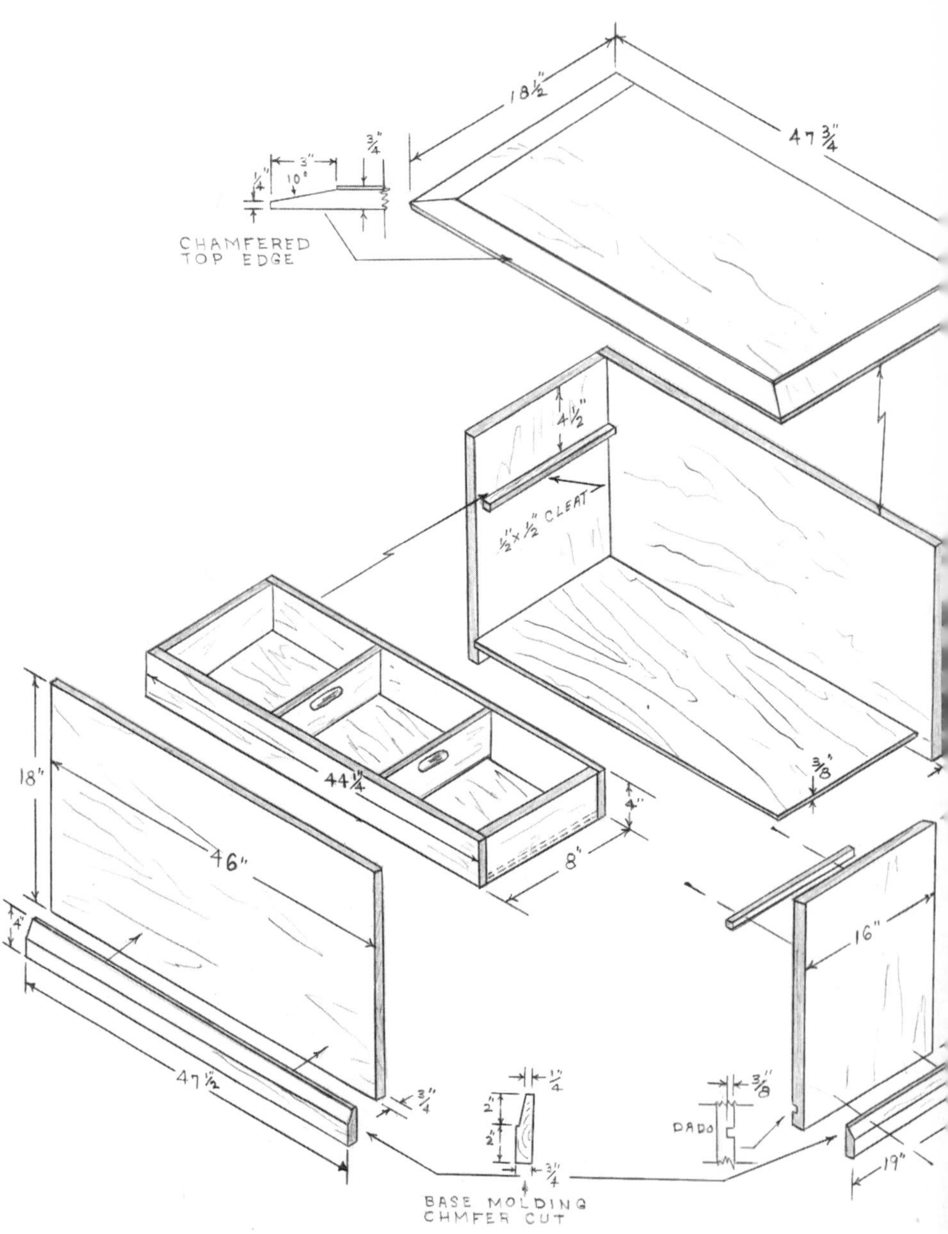

18½"
47¾"
3"
¾"
¼"
10°
CHAMFERED
TOP EDGE
4½"
½" x ½" CLEAT
44¼
4"
8"
3/8"
18"
46"
4"
47½
¾"
16"
19"
¼
2"
2"
¾
3/8
DADO
BASE MOLDING
CHMFER CUT

*Note:* As an alternative to fastening the corners together using the simple butt joints shown in the drawing, a miter or rabbet joint may be used.

*Note:* Glue blocks should be attached to the base to support the bottom.

4. To assemble the shelf cut a ⅜-inch by ⅜-inch dado into each end piece approximately ½ inch up from the bottom.

   Assemble the shelf by gluing the ⅜-inch bottom into the dadoes then nailing the four sides together with #6 finishing nails.

   Support the bottom with ½-inch by ½-inch glue blocks.

   Drill two ¾-inch-diameter finger holes into each of the two divider boards, then fasten the dividers into the shelf with #6 finishing nails.

5. Attach the two ½-inch by ½-inch cleats to the ends with glue and 1-inch brads.

6. Set the table saw to approximately a 15-degree angle and a 2-inch depth to make the chamfered design for the base molding.

7. To fasten the base molding to the chest, cut the corners at a 45-degree angle, then attach the molding with glue and #6 finishing nails.

   *Note:* You may use 4-inch clamshell molding for the base. This can be purchased at your local lumber yard.

8. To make the chamfered design on the top edge, set the table saw to approximately a 10-degree angle and to a 3-inch depth, then make the chamfered cut on the front and two ends of the top piece.

   Fasten the top to the back with 3-inch surface H-hinges or with special lid hinges. A lid support should also be fastened to the top and inner side.

9. Set and fill all holes, scrape off all traces of glue, then sand the chest to a smooth surface. The final sanding should be done in a straight line with the grain by hand sanding.
10. Select and apply the colored stain of your choice. When dry, apply several coats of clear spray lacquer, rubbing lightly between coats with fine pumice and oil or fine steel wool. Protect the finish with a coat of paste wax and buffing.

    *Note:* On my hope chest I used ¼-inch by ¾-inch strips to give the appearance of doors and to conceal the corners.

## Project 13

# TWO-DRAWER CHEST

### Materials Required

*Wood of Your Choice*

**1 top** ¾″ × 16½″ × 25½″
**2 sides** ¾″ × 15″ × 23¼″
**2 shelf and drawer divider panels** ¾″ × 14¾″ × 23¼″
**1 top drawer rail** ¾″ × 1″ × 22½″
**1 front baseboard skirt** ¾″ × 5½″ × 24″
**1 back** ¼″ × 23¼″ × 23¼″
**1 large drawer unit** (overall size) 10¾″ × 14¼″ × 23″ with a ⅜″ lip around the front
**1 small drawer unit** (overall size) 4¼″ × 14¼″ × 23″ with a ⅜″ lip around the front
**hardware** #10, 1½″ flathead wood screws, furniture buttons, or dowel plugs of your choice

*Note:* Refer to procedure on construction of night table drawers.

### Procedures

1. Lay out and cut all pieces to the suggested size and design.

2. Lay out and cut a ¼-inch by ⅜-inch rabbet on the back inside edge of each side piece. These cuts are made to receive the ¼-inch back.

3. Lay out and cut the two ⅜-inch by ¾-inch dado joints into each side piece, then proceed to insert the shelf and drawer divider into the dadoes by gluing and clamping with bar clamps.

88

# Two-Drawer Chest

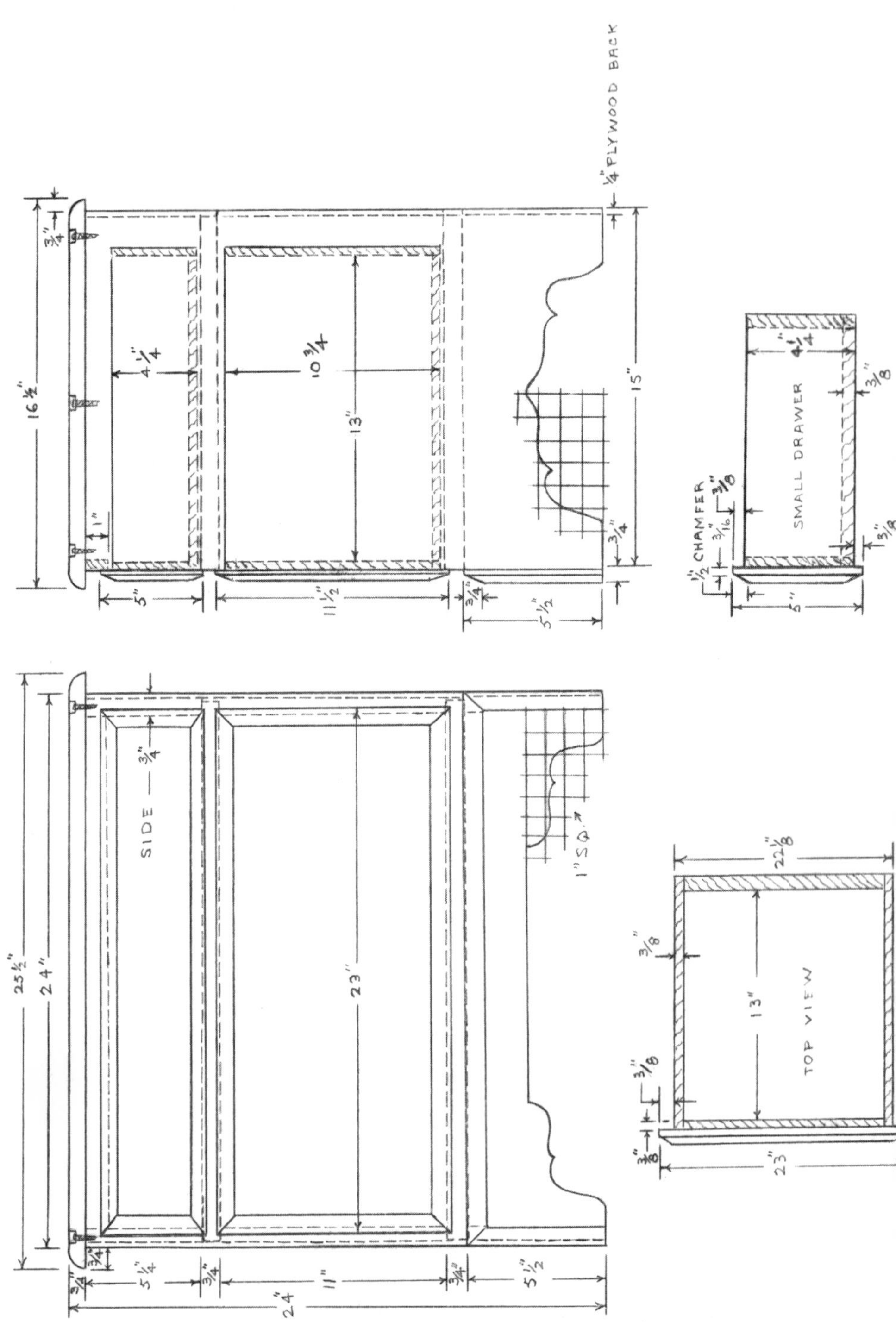

*Note:* The divider panels can be made from particle board with only the exposed front nosed with the wood used in the rest of the chest. This will save on cost. Be sure to check for squareness.

4. Attach the ¼-inch back to the precut rabbet with ¾-inch finishing brads.
5. Locate and drill six ⅜-inch counterbored pilot holes through the top and into the side pieces. Continue to fasten the top to the chest assembly by using #10, 1½-inch flathead wood screws.
6. Fasten the ¾-inch by 1-inch drawer rail below the surface of the top by gluing and clamping. The rail fills the space between the top and the small drawer.
7. Lay out and cut a ¼-inch by ¾-inch chamfer on the three edges of the base skirt, then fasten the skirt to the front edge of the sides by gluing and nailing into position with #6 finishing nails.
8. Construct the two drawer units to the overall size given in the drawing. Be sure to cut a ⅜-inch by ⅜-inch rabbet on the inside edge of the front pieces to allow a lip so the front can overlap the cabinet frame. Refer to drawer construction hints at the front of the book.
9. Shape the quarter-round cut on the top piece using a router and ¼-inch round bit or by hand with a jack plane and file.
10. Scrape off all traces of glue, set and fill all nail holes, then sand the entire chest to a smooth surface. When completely sanded, fill the ⅜-inch counterbored screw holes with ⅜-inch furniture buttons or ⅜-inch dowel plugs.

11. Select and apply the colored stain of your choice. When dry, apply several coats of clear polyurethane finish. Rub lightly between coats of finish with #320 emery cloth or pumice and oil. Protect the finish with a coat of paste wax and buffing.
12. Install your choice of cabinet hardware.

## Project 14

# COFFEE TABLE

### Materials Required

*Wood of Your Choice*

**1 top** 1″ × 20″ × 42″
**4 legs** 2¼″ × 2¼″ × 16″
**2 aprons** ¾″ × 2½″ × 15¼″
**2 aprons** ¾″ × 2½″ × 38″
**8 glue blocks** ¾″ × 2″ × 4″
**wood screws** #10, 2¾″ flathead

### Procedures

1. Edge-glue enough stock to reach the required width for the top.

2. Lay out and cut the top to the size indicated in the drawing. Use a router to shape the edge to a quarter-round or chamfered design.

3. Lay out and cut the four pieces for the apron. Cut a ⅜-inch by ¾-inch rabbet at the ends of the front and back pieces of the apron.

   Fasten the four pieces together using glue and #6 finishing nails.

   Secure the apron to the underface of the top by gluing and clamping. Reinforce by drilling counterbored pilot screw holes and fastening #10, 2¾-inch flathead screws into the underface of the top.

4. Laminate enough stock to reach the 2¼-inch by 2¼-inch thickness for the legs, then lay out and cut the taper of the four legs to the dimensions given in the drawing.

# Coffee Table

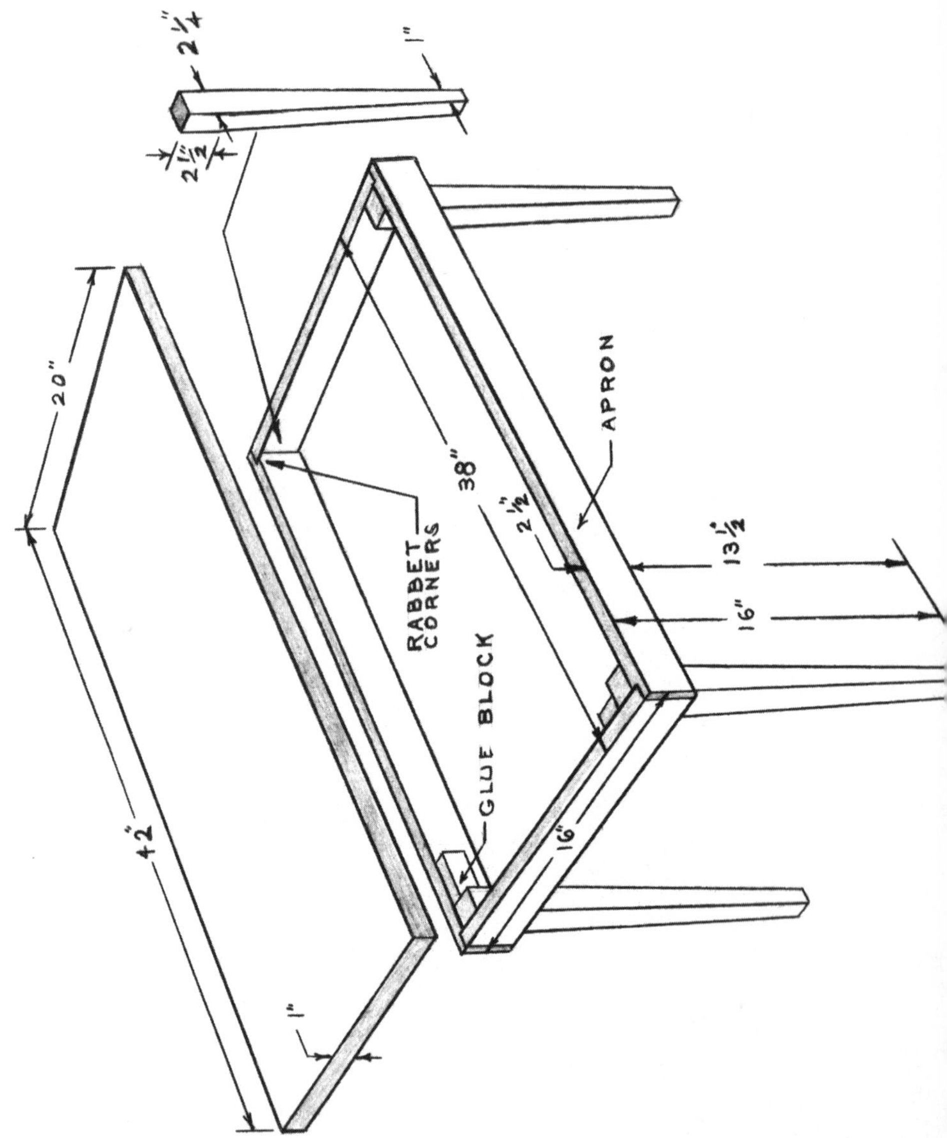

5. Secure the legs to the inner corners of the apron by using #10, 2¾-inch screws fastened through four countersunk pilot holes. Reinforce the legs to the inner corners of the apron by fastening glue blocks.
6. Set and fill all nail holes. Use wood filler if the wood is open grain, then sand the table to a final smooth finish.
7. Select and apply the stain of your choice. When dry, apply four coats of polyurethane finish, rubbing between coats with 320 emery paper. On the final coat rub with pumice and lemon oil. Preserve the finish with an occasional coat of lemon oil.

## Project 15

# BREAD OR PIE BOARD

### Materials Required

*Wood of Your Choice*

**1 handle** ¾″ × 3¼″ × 11″
**1 board** ½″ × 10″ × 16″
**2 wood screws** #9, 1¼″ flathead
**2 dowel pins** ⅜″ dia. × 1½″

### Procedures

1. Glue up and clamp the stock to the required width, then lay out and cut both pieces to the size and shape indicated. The edges of the handle and board should be rounded with a half-round file and sandpaper.

2. Locate and drill two countersunk pilot holes through the bottom of the board and into the handle, then fasten the handle to the board with #9, 1¼-inch flathead wood screws. An optional process for fastening the handle is to bore two ⅜-inch-diameter holes through the handle and board then inserting two ⅜-inch dowel pins into the holes with glue.

3. Locate and drill a 3⁄16-inch-diameter hole through the handle for hanging the board.

4. Scrape off all traces of glue, then sand the project to a smooth surface.

5. Apply several coats of clear finishing oil, rubbing between coats with fine steel wool or pumice and oil.

# Bread or Pie Board

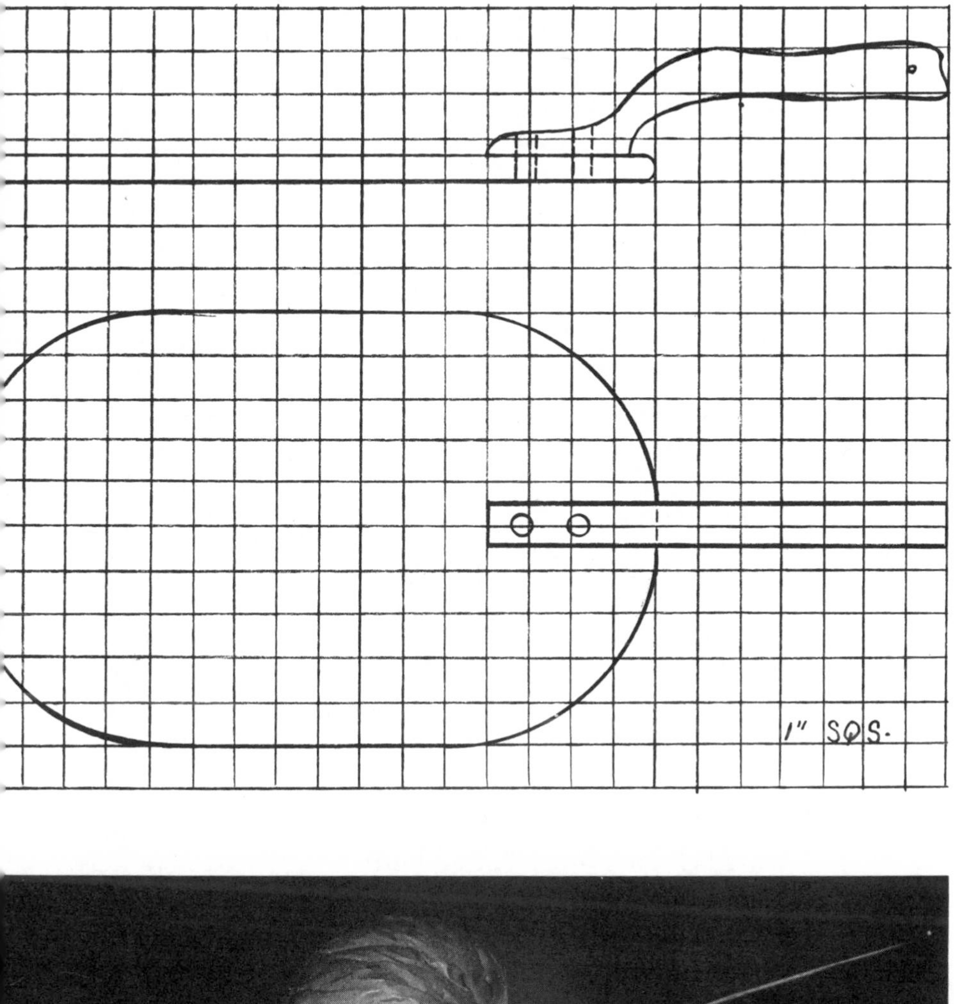

1" SQS.

## Project 16

# FOOT REST

### Materials Required

*Wood of Your Choice*

**2 sides** 1″ × 18″ × 18¼″
**2 rails** ¾″ × 3½″ × 22″
**1 seat panel** ½″ × 16″ × 22″
**2 cleats** ¾″ × 2″ × 16″
**6 wood screws** #9, 1¼″ flathead
**8 wood screws** #12, 1½″ flathead
**1 loose cushion** Available at your local department store.
**1 yard of fabric**

### Procedures

1. Glue up enough stock to meet the required width for the two sides, then proceed to lay out and cut out all of the pieces to the recommended size and design.
2. Use a ¼-inch round bit in a router or a spoke shave and file to make the ½-round cut on the edge of the side pieces. The outside edge of the rails should be ¼ round.
3. Drill four countersunk pilot screw holes through each cleat and into the side pieces.

   Attach the cleats to the side pieces with #12, 1½-inch flathead wood screws. Attach the ½-inch-thick seat panel on top of the two cleats by drilling three countersunk pilot screw holes through the wood and into each cleat support. Insert #9, 1¼-inch flathead wood screws.

# Foot Rest

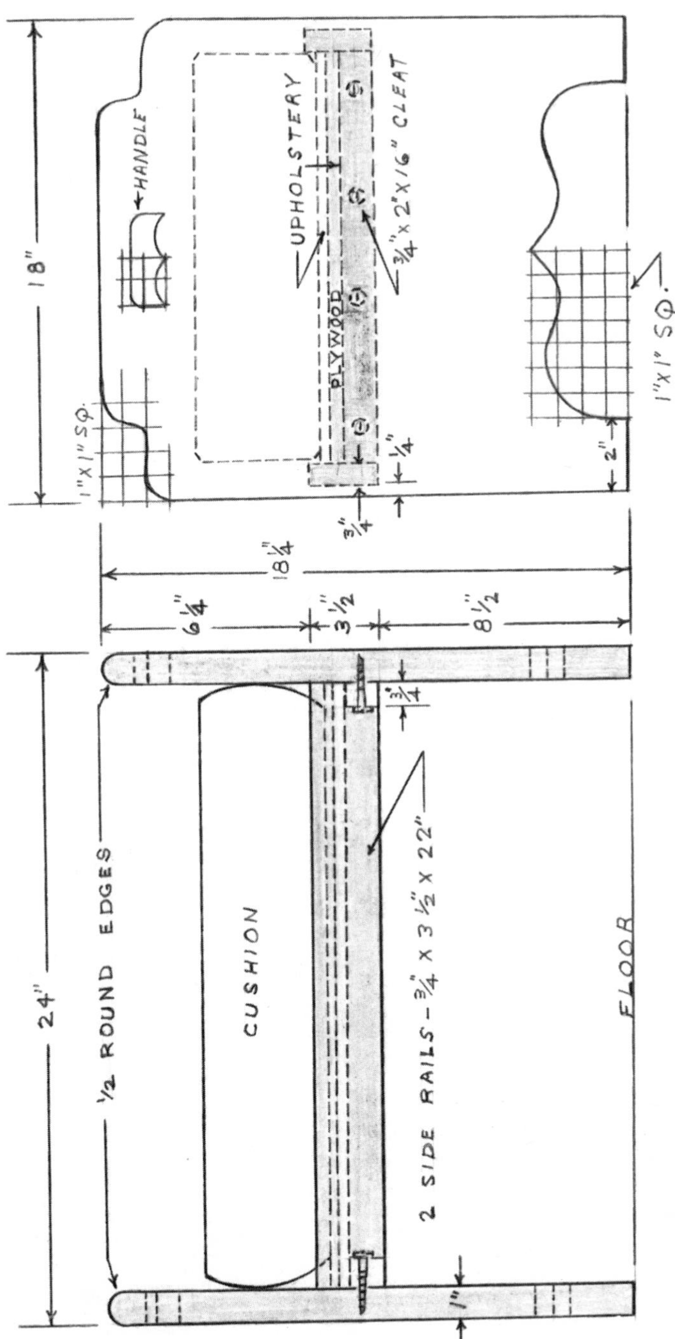
18"
HANDLE
UPHOLSTERY
PLYWOOD
3/4" x 2" x 16" CLEAT
1" x 1" SQ.
1" x 1" SQ.
1/4
3/4"
2"
18 1/4"
6 1/4"
3 1/2"
8 1/2"
24"
1/2 ROUND EDGES
CUSHION
2 SIDE RAILS - 3/4" x 3 1/2" x 22"
3/4"
1"
FLOOR

Finish the assembly by fastening the two rails to the outside edge of the wood and cleats. Use glue and #6 finishing nails to hold the rails.

4. Set and fill all nail holes, then proceed to sand the foot rest to a smooth, scratch-free surface.
5. Select and apply the colored stain of your choice. When dry, apply a minimum of four coats of finishing oil, rubbing between dry coats with #320 grit emery cloth or fine steel wool. Protect the final finish with a coat of paste wax and buffing.

   *Note:* The upholstery over the plywood is an option. If used, it should be attached with glue and upholstery tacks.

## Project 17

# NIGHT CHEST

### Materials Required

*Wood of Your Choice*

**1 top** ¾″ × 14¾″ × 19¾″
**2 sides** ¾″ × 15¼″ × 25¼″
**3 shelf and drawer dividers** ¾″ × 15″ × 17″
**1 top trim piece** ¾″ × 1½″ × 16¼″
**1 baseboard skirt** ¾″ × 5″ × 17¾″
**1 back** ¼″ × 17″ × 25¼″
**2 drawers** (overall size) 3¾″ × 14¼″ × 16¾″ with a ⅜″ lip around the front

### Procedures

1. Lay out and cut all pieces to the recommended size and shape. Lay out and cut a ¼-inch by ⅜-inch rabbet cut on the back inside edge of each side piece. These cuts are made to receive the ¼- inch back.
2. Lay out and cut the three dado joints into each side piece.

   Proceed to assemble the two drawer dividers and shelf into the precut joints by gluing and clamping with bar clamps.

   Fasten the ¼-inch back into the precut rabbet joints using 1-inch finishing brads.

   *Note:* Before fastening the back in place make sure the assembly is square.
3. Locate and drill six ⅜-inch counterbored pilot holes through the top and into the side pieces.

   Fasten the top to the chest assembly with #10, 1½-inch flathead wood screws.

# Night Chest

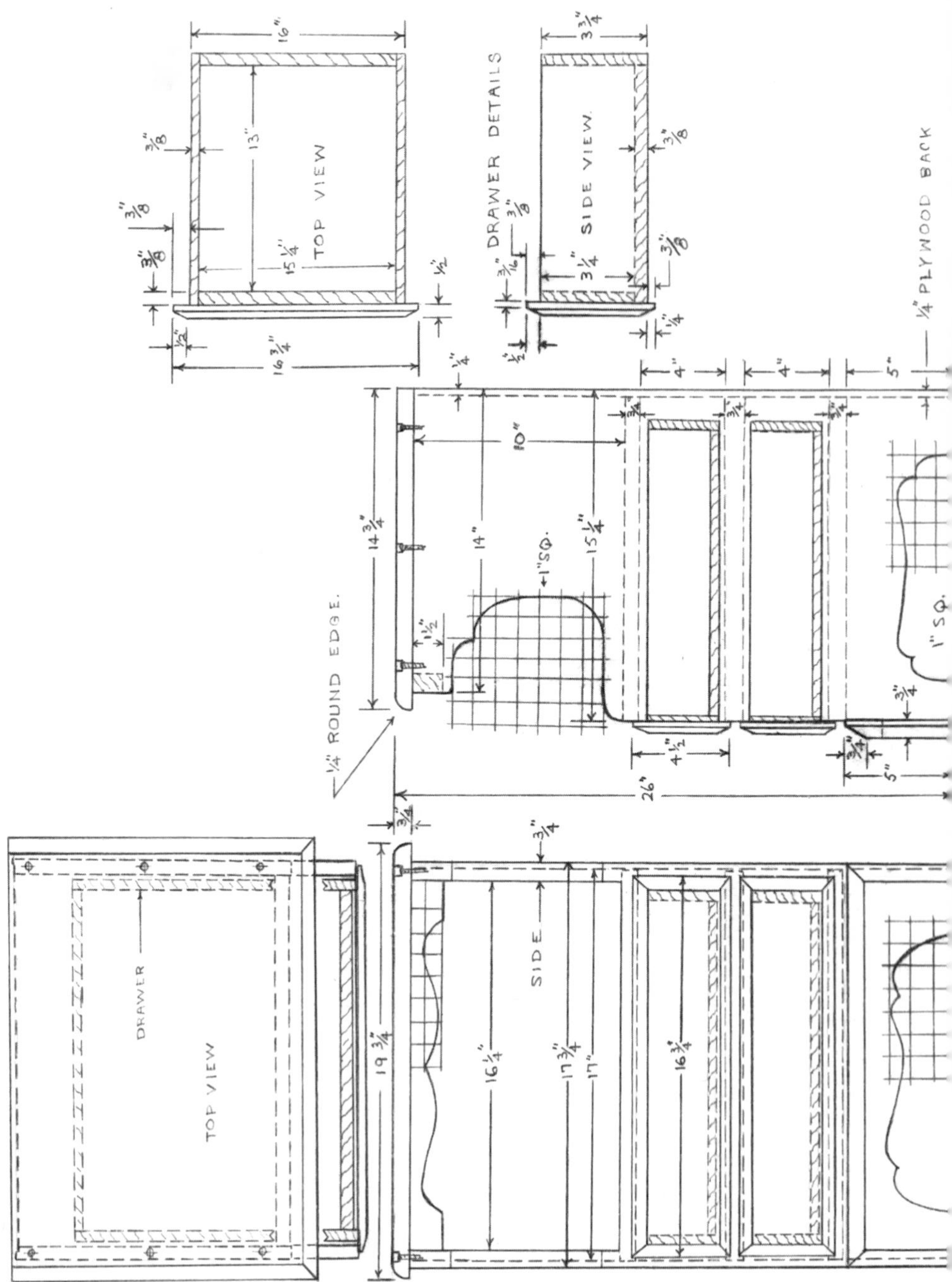
TOP VIEW
DRAWER DETAILS
SIDE VIEW
1/4" PLYWOOD BACK
1/4" ROUND EDGE
1" SQ.
1" SQ.
SIDE
DRAWER
TOP VIEW

GIFT PROJECTS
Running

Fill the counterbored holes with ⅜-inch dowel plugs or ⅜-inch furniture buttons.

Attach the 1½-inch trim to the bottom of the top and to the two sides by gluing and clamping.

4. Lay out and cut a ¼-inch by ¾-inch chamfer on the exposed three edges of the base skirt, then proceed to fasten the skirt to the sides by gluing and nailing with #6 finishing nails.

   *Note:* The chamfer can be cut by tilting the arbor of a table saw to approximately 30 degrees, or it can be cut by hand with the use of a jack plane.

5. Construct the two drawer units to the overall size given in the drawing. Be sure to cut a ⅜-inch by ⅜-inch rabbet cut on the inside edge of the front to allow the drawer to overlap the frame of the chest. Refer to drawer construction hints at the front of the book. Cut a chamfer on the four outside edges of the drawer.

6. Shape the rounded edge on the top piece by using a hand plane and file or a router with a quarter-round bit.

7. Use ⅜-inch dowel plugs or ⅜-inch furniture buttons to plug screw holes.

   Set and fill nail holes, then proceed to sand the project to a smooth surface.

8. Select and apply the colored stain of your choice. When dry, apply several coats of clear polyurethane finish. Rub lightly between coats with #320 emery cloth or pumice and oil. Protect the finish with a coat of paste wax.

9. Install your choice of cabinet hardware to each drawer.

## Project 18

# MAGAZINE RACK

### Materials Required

*Wood of Your Choice*

**12 dowels** 3″ × 12″
**2 dowels** 3/8″ × 6″
**2 dowels** 1½″ × 17½″
**1 base** ¾″ × 5½″ × 17½″

### Procedures

1. Lay out and cut all parts to the suggested size.
2. Lay out the location and bore all holes in the base. The twelve 3/8-inch-diameter holes bored in the base should be positioned at a 70-degree angle. To assure the correct angles set a sliding T-bevel to 70 degrees to check the angle of the bit cut.
3. Lay out the location and drill all the 3/8-inch-diameter holes ¾ inch into the two 1½-inch-diameter dowels. Make sure all of the holes are bored at a 90-degree angle.
4. Fasten all of the pieces into the predrilled holes.
5. Clean off all traces of glue and continue to sand the project to a smooth surface.
6. Select and apply the colored stain or paint of your choice. When dry, apply several coats of clear lacquer or polyurethane finish. Rub between each coat with fine steel wool or pumice and oil. Preserve the finish with lemon oil.

# Magazine Rack

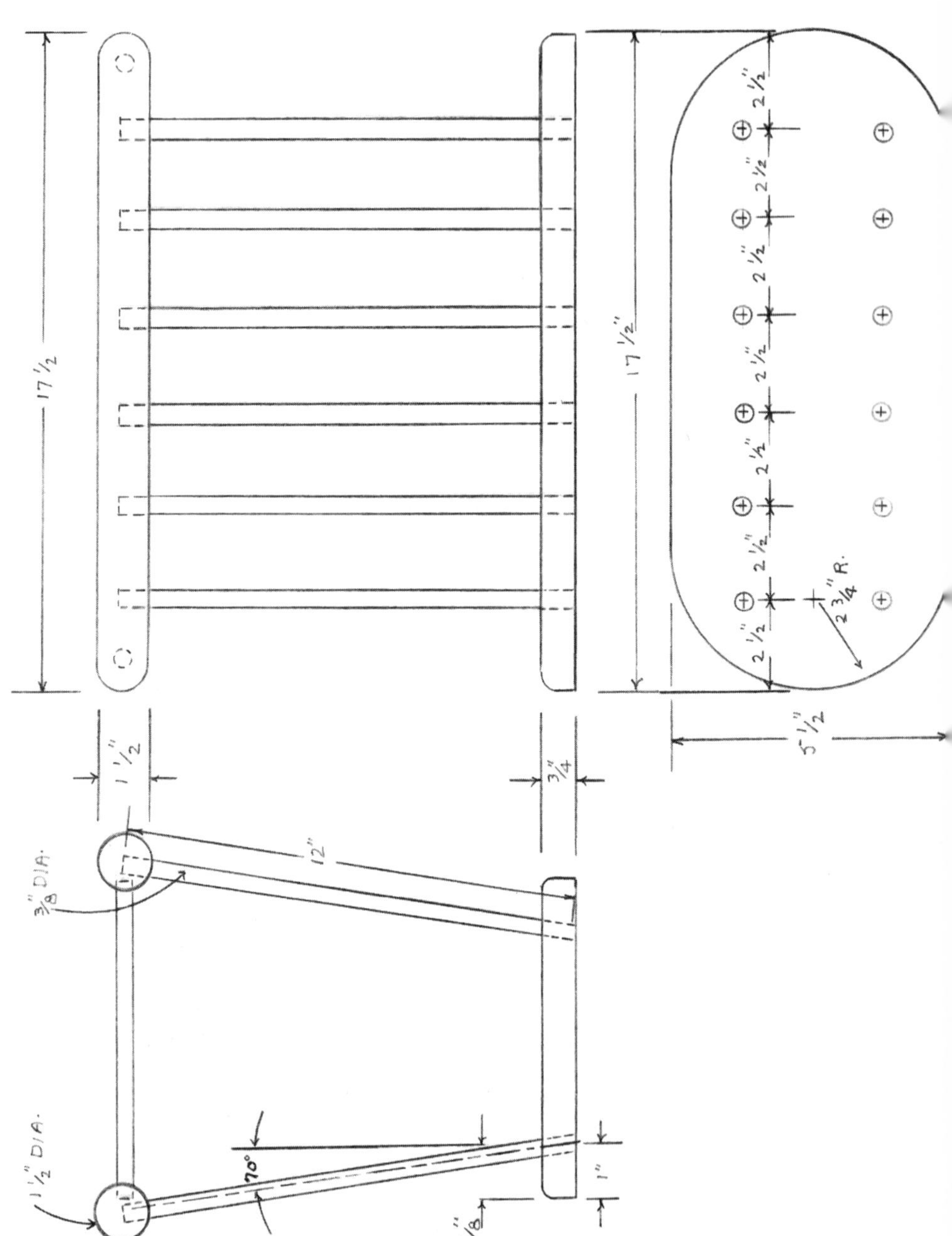

## Project 19

# TOWEL DRYING STAND

### Materials Required

*Wood of Your Choice*

**4 side design pieces** 1¼″ × 3¾″ × 14″
**6 cross dowels** ¾″ dia. × 30½″
**4 vertical dowels** ¾″ dia. × 27½″

### Procedures

1. Lay out and cut all of the stock to the size and design given in the drawing. It may be necessary to laminate stock together to reach the required thickness for the four side pieces.

2. Lay out the location, then proceed to bore the three ¾-inch-diameter holes to a ½-inch depth in each side piece to receive the horizontal dowels.

   Lay out and bore the two ¾-inch-diameter holes in each side piece to receive the vertical dowels. Be careful in aligning the holes to one another. Use a spade bit to bore holes.

3. Round over the edges on the side design pieces with a half-round file and sandpaper. A ⅛-inch rounding is attractive.

4. Insert the four vertical dowels into the four side design pieces. Use glue for securing into position.

   When the glue has dried, complete the assembly by gluing and clamping the six cross dowels into the prebored ¾-inch-diameter holes.

# Towel Drying Stand

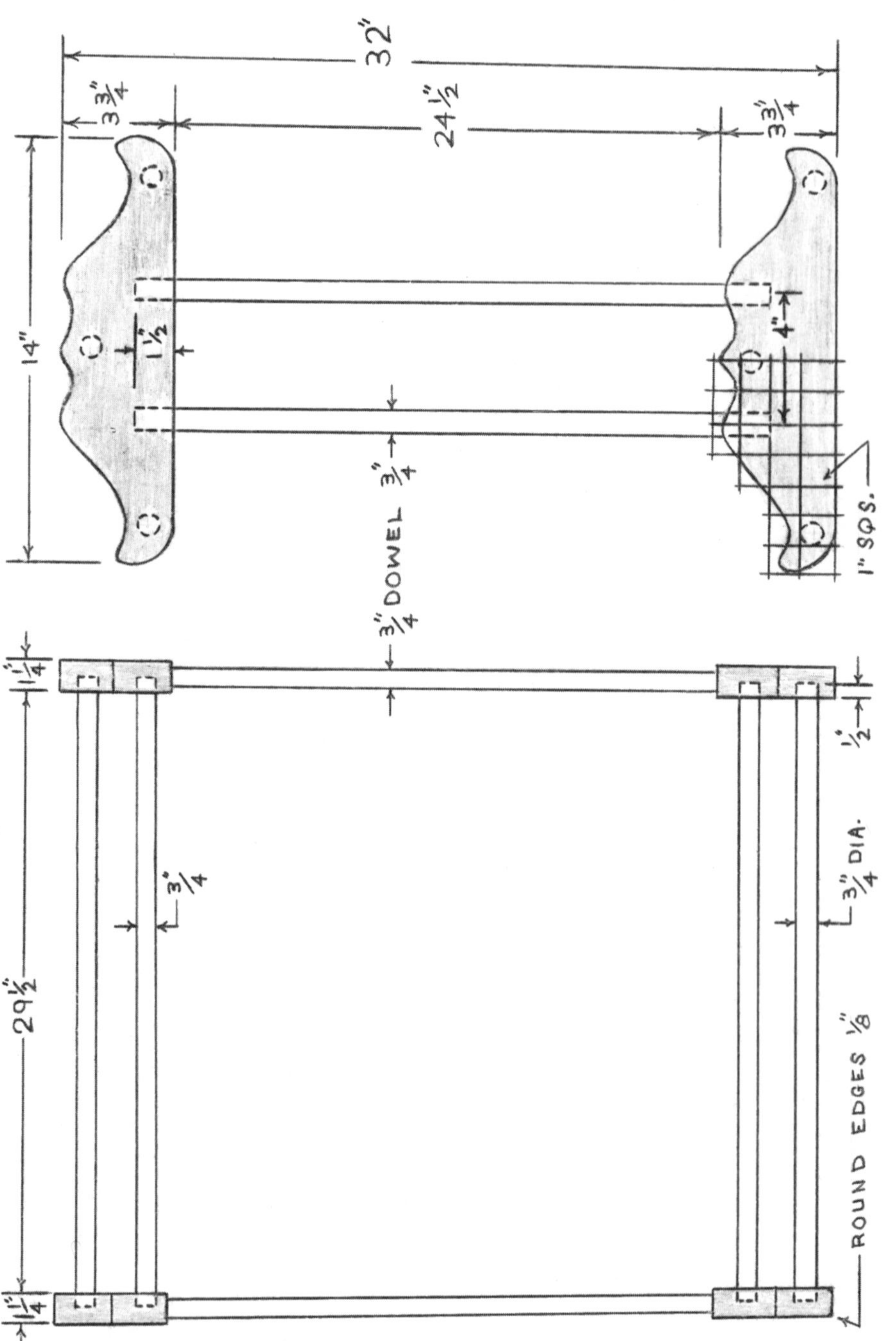

*Note:* When the cross dowels are being glued in place, use a framing square to check the stand for squareness.

5. Scrape all traces of glue, then sand the entire project to a smooth surface.
6. Select and apply the colored paint or stain of your choice. When completely dry, apply several coats of spar varnish, rubbing between coats with 320 emery paper or fine steel wool. Preserve the finish with an occasional coat of paste wax.

   *Note:* Spar varnish is used for its moisture-repelling additives.

## Project 20

# WALL PLANTER

### Materials Required

*Wood of Your Choice*

**1 back** ¾″ × 10″ × 24″
**2 sides** ½″ × 5½″ × 6″
**1 front** ½″ × 5½″ × 9″
**1 bottom** ½″ × 5¾″ × 9″
**1 mirror** ⅛″ × 6½″ × 14½″ Available at your local glass dealer.

### Procedures

1. Lay out and cut the back to the size and design indicated in the drawing. The inner cutout for the mirror can be made by boring a ½-inch hole within the pattern layout and then inserting the blade of a coping saw or sabre saw to make the cut.
2. File the inner edge of the mirror cutout smooth, then proceed to make a ¼-inch by ⅜-inch rabbet cut into the back of the mirror cutout. Use a rabbet bit inserted into an electric router.

   *Note:* A more simple procedure to fasten the ⅛-inch mirror to the back is to hold it in place with tacks spaced approximately 3 inches apart. With this method the rabbet cut is not required to hold the ⅛-inch mirror.
3. Lay out and cut the two sides and front to the suggested size. The bottom and top edge of these pieces should be planed or cut to approximately 15 inches so they will fan out. Use a sliding T-bevel and protractor to check the

# Wall Planter

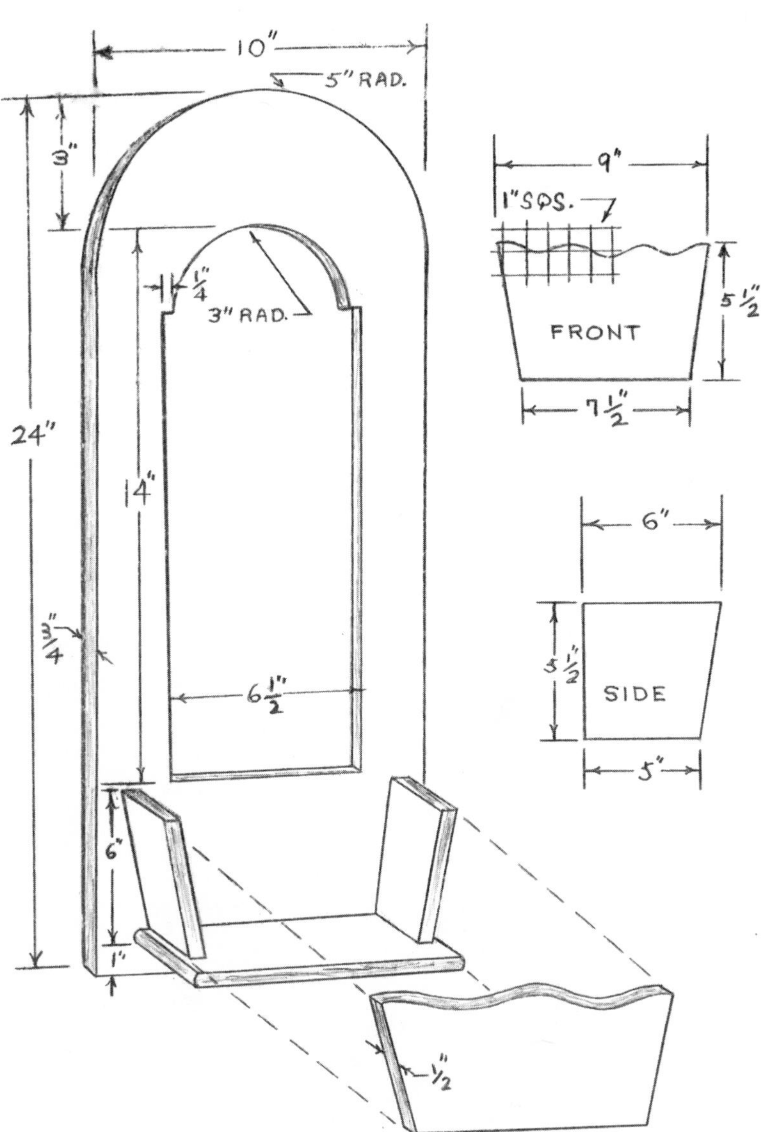

15-degree angle. Lay out and cut the bottom ⅛ inch on all pieces.

4. Fasten the four pieces to the back by drilling counterbored holes then installing #8, 1-inch screws.
5. Use hardwood plugs to cover holes. Scrape off all traces of glue, then sand the planter to a smooth surface.
6. Select and apply the stain of your choice. When dry, apply three coats of polyurethane, rubbing between dry coats with 320 emery paper or fine steel wool. Preserve the finish with an occasional coat of lemon oil.
7. If the rabbet cut is made on the back of the mirror cutout, have your local glass dealer cut and set the mirror into the ¼-inch by ⅜-inch rabbet cut.

## Project 21

# CANDLE SCONCE

### Materials Required

*Wood of Your Choice*

**1 back** ¾″ × 5″ × 18″
**1 bracket** ¾″ × 2½″ × 3¾″
**1 disk** ¾″ × 2¾″ dia.
**1 disk** ¾″ × 2⅜″ dia.

*Note:* A metal inset or candle cup can be purchased at your local craft supply store.

*Note:* The disks are made to hold a 2½-inch-diameter glass chimney. The glass chimney can be purchased at your local hardware or craft supply store.

### Procedures

**1.** Lay out and cut all of the pieces to the suggested size and design.

**2.** If a candle cup is not used, bore a ⅞-inch hole through the center of the 2⅜-inch-diameter disk to hold the candle. A ⅞-inch-diameter metal inset can be inserted into the hole. If a candle cup is used, glue and clamp it to the top disk.

**3.** Glue and clamp the small disk to the large disk, making sure the small disk is centered over the larger one.

**4.** Assemble the two disks to the bracket by gluing and clamping.

# Candle Sconce

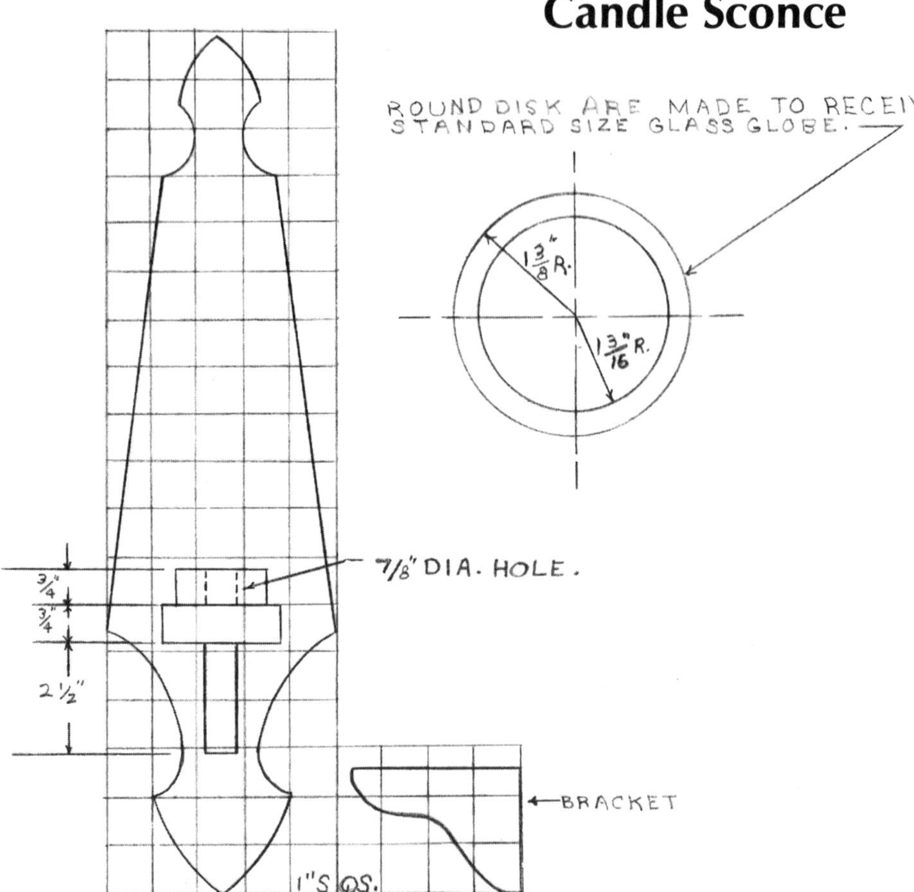


5. Fasten the bracket and disk assembly to the back by drilling a countersunk pilot hole through the back and into the bracket, then attach with two #8, 1½-inch flathead wood screws.

6. Scrape off all traces of glue, then sand the sconce to a smooth surface.

7. Select and apply the stain of your choice. When dry, apply several coats of clear finishing oil, rubbing between each coat with fine steel wool or pumice and oil. Protect the finish with a coat of paste wax and buff.

## Project 22

# WALL CUPBOARD

### Materials Required

*Wood of Your Choice*

**1 top design piece** ¾″ × 2¾″ × 24″
**1 bottom design piece** ¾″ × 1″ × 24″
**2 shelves** ¾″ × 11½″ × 22½″
**1 inner top frame** ¾″ × 11½″ × 22½″
**1 inner bottom frame** ¾″ × 11½″ × 22½″
**1 outer top** ¾″ × 12½″ × 24″
**1 outer bottom** ¾″ × 12½″ × 24″
**2 doors** ¾″ × 11 15/16″ × 26⅜″
**2 door handles** ½″ × 1¼″ × 1″
**2 dowels for handles** ¼″ dia. × 1½″
**1 back (optional)** ¼″ × 23¼″ × 27¼″
**butt hinges** (2″) or **wooden dowel pins** (1¼″) **(optional)**

### Procedures

1. Edge-glue enough stock to achieve the required width for all pieces.
2. Lay out and cut all pieces to the size and design shown in the drawing.
3. Cut a ¼-inch by ⅜-inch rabbet on the back inner edge of the four sides. This can be done with a table saw or a router and rabbet bit. This rabbet cut is to receive the back when the sides are fastened.
4. Lay out and fasten the two shelves along with the inner top and bottom frames to the sides of the cabinet. Use #6 finishing nails.

   Attach the ¼-inch back to the precut rabbet with 1-inch finishing brads. Be sure the cabinet is square before attaching the back.

# Wall Cupboard

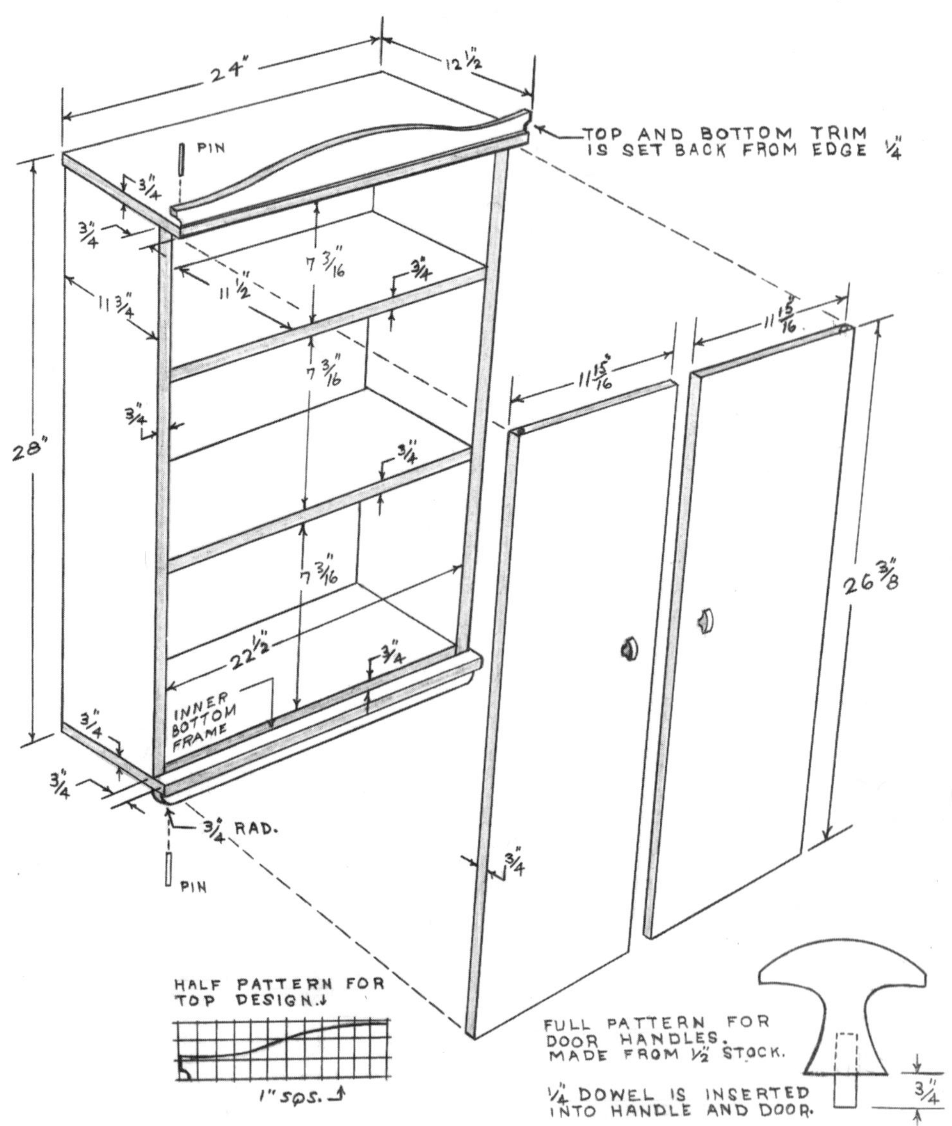

*Note:* The shelves are secured to the sides with the simple butt joint, but you may prefer to use dadoes or more elaborate joints.

5. Attach the top and bottom by gluing and clamping to the face of the inner top and bottom frames. The front of both pieces should overhang 1 inch to receive the doors.

6. In making pivotal dowel hinges, begin by using clamps to hold both doors in position to the front of the cabinet, then drill through the top and bottom overhangs on into the doors with a ¼-inch-diameter drill. The holes should be located approximately ⅝ inch from the outside edge of each overhang. To prevent the doors from rubbing against the bottom use a small nylon or metal washer over the bottom dowel pivot pin. Insert the ¼-inch-diameter dowel pins through the overhangs and into the doors.

   *Note:* Use soap or wax on the dowel pins for less noise and friction when opening and closing. To allow for a smooth swing of the doors use a hand plane to cut a chamfer on the corners of the inner edge. This will prevent the door from hitting the edge when opening.

7. Make two door handles to the pattern suggested. Drill a ¼-inch-diameter hole approximately ¾ inch through the bottom of the handle, then glue in a ¼-inch by 1½-inch dowel.

   Locate the door handles to approximately the center of the doors and 4 inches apart.

8. Finish the cabinet assembly by attaching the top and bottom trim by gluing and clamping.

9. Select and apply the colored stain of your choice. When completely dry, apply a minimum of four coats of clear finishing oil, rubbing between coats with #320 emery paper. For the final rubbing use pumice and lemon oil. Preserve the finish with an occasional coat of lemon oil.

## Project 23

# BEDSIDE WALL SHELF

### Materials Required

*Wood of Your Choice*

**1 top** 5/8″ × 12 1/2″ × 30″
**2 sides** 5/8″ × 12″ × 18″
**1 shelf** 5/8″ × 12″ × 26″
**1 back** 5/8″ × 4″ × 26″

*Drawer:*
**1 front** 1/2″ × 4″ × 25 7/8″
**1 back** 1/2″ × 4″ × 25 7/8″
**2 sides** 1/2″ × 4″ × 9″
**2 runners** 3/8″ × 3/8″ × 9″
**1 bottom** 1/4″ × 9″ × 25 3/8″
**drawer pulls of your choice**

### Procedures

1. Lay out and cut the top to size.

   Round over the three exposed edges or make a cove cut by the use of the electric router and cove bit.

   Lay out and drill the four counterbored pilot holes for the screws to secure the top to the sides.
2. Cut the side to the suggested pattern, then locate and drill two counterbored pilot holes in each piece so that screws can secure the shelf to the sides.
3. Cut the shelf to the correct size. The exposed edge can be rounded over, or a cove cut can be used if desired.

# Bedside Wall Shelf

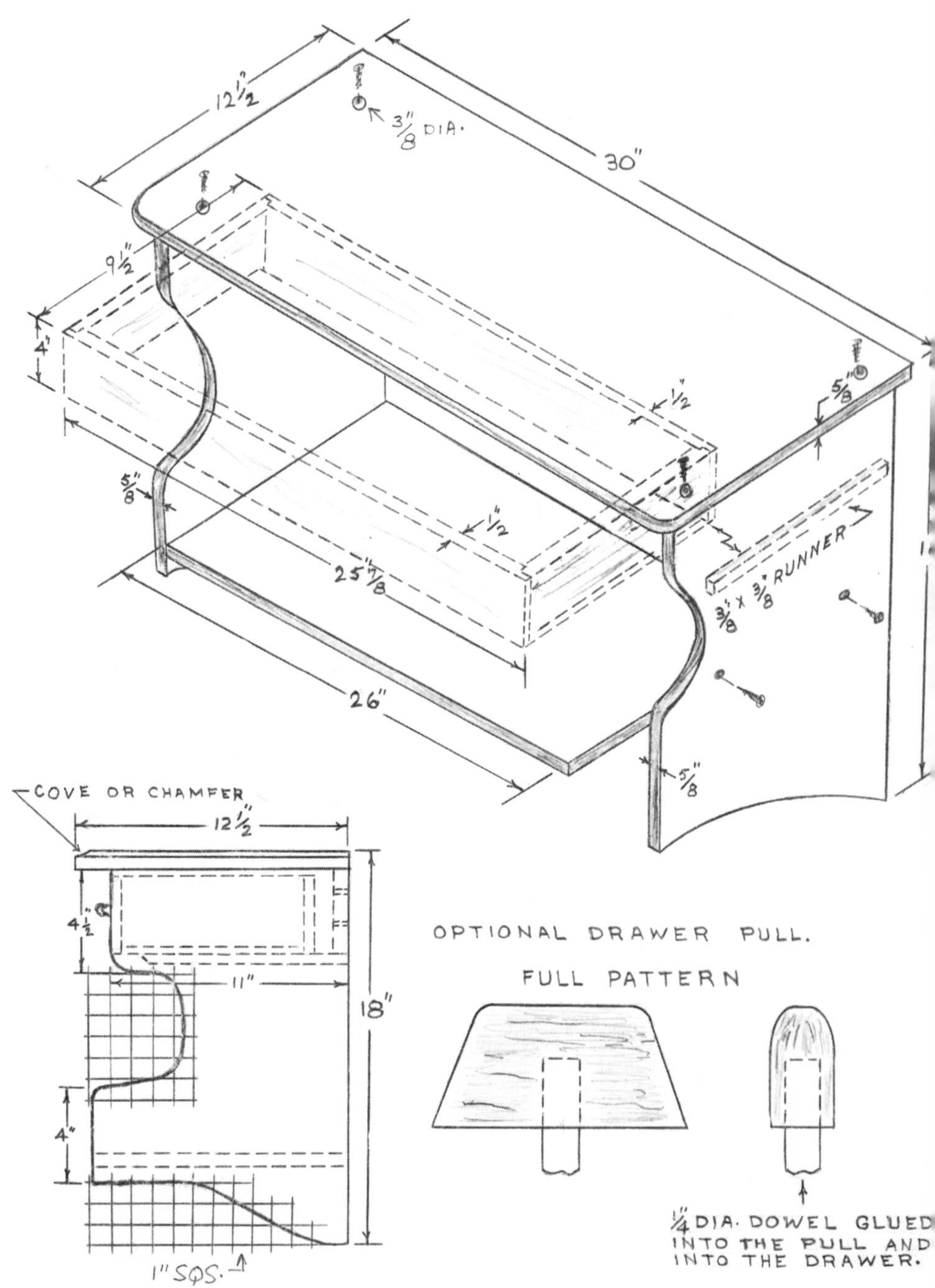

Young Lives at Stake

Assemble the two sides of the shelf by screwing #10, 1½-inch flathead wood screws through the predrilled pilot holes, then attach the top to the top edge of the sides in the same manner.

4. Construct the 4-inch by 9½-inch by 25⅞-inch drawer unit by making ¼-inch by ½-inch rabbet joints on both ends of the front and back pieces.

   *Note:* A simpler method is to butt the sides together in order to fasten the drawer. (Refer to the section on drawer construction in the front part of the book.)
5. Glue the two ⅜-inch by ⅜-inch by 9-inch runners approximately 4⅛ inches down from the top onto the side pieces.
6. If making the optional drawer pulls, cut them out to the full pattern given, then drill a ¼-inch-diameter hole ¾-inch deep into the pull and then into the desired location in the drawer. Fasten the dowel to the drawer pull and drawer with glue.
7. Plug all ⅜-inch holes with ⅜-inch furniture buttons or dowel plugs.

   Sand the shelf to a smooth surface, then apply several coats of polyurethane or lacquer. Rub between coats lightly with fine steel wool. Preserve the finish with an occasional coat of paste wax or lemon oil.

## Project 24

# WALL SHELF

### Materials Required

*Wood of Your Choice*

**1 shelf** ¾″ × 9½″ × 30″
**2 brackets** ¾″ × 7″ × 9″
**1 back** ¾″ × 2″ × 30″
**4 wood screws** #8, 1½″ flathead

### Procedures

1. Lay out and cut all pieces to the suggested size and design.
2. Lay out and cut the ¾-inch by 2-inch rabbets on the back of the bracket pieces.
3. Attach the back piece to the cut rabbets in the brackets by drilling pilot and countersink holes then fastening two wood screws into each bracket. Fasten the top piece to the brackets and back by gluing and clamping.
4. Drill two ³⁄₁₆-inch-diameter holes exactly 16 inches apart into the back pieces. The holes are located so they are aligned to the studs in a wall for hanging the shelf.
5. Clean off all traces of glue, set and fill and nail holes, then sand the shelf to a smooth surface.
6. Select and apply the colored stain or paint of your choice. When dry, apply several coats of clear polyurethane, rubbing lightly between each coat with fine steel wool or pumice and oil. Protect the finish by applying paste wax and buffing.

# Wall Shelf

TWO HOLES 16" ON CENTER
5½"
9½"
2½"
7"
27"
¾"
2"
9¾"
3½"
18½"
3½"
¾
¾
1" SQS.
CUT A 3/4" X 2" RABBET FROM BACK OF BRACKET.
ALL STOCK IS 3/4" THICK.

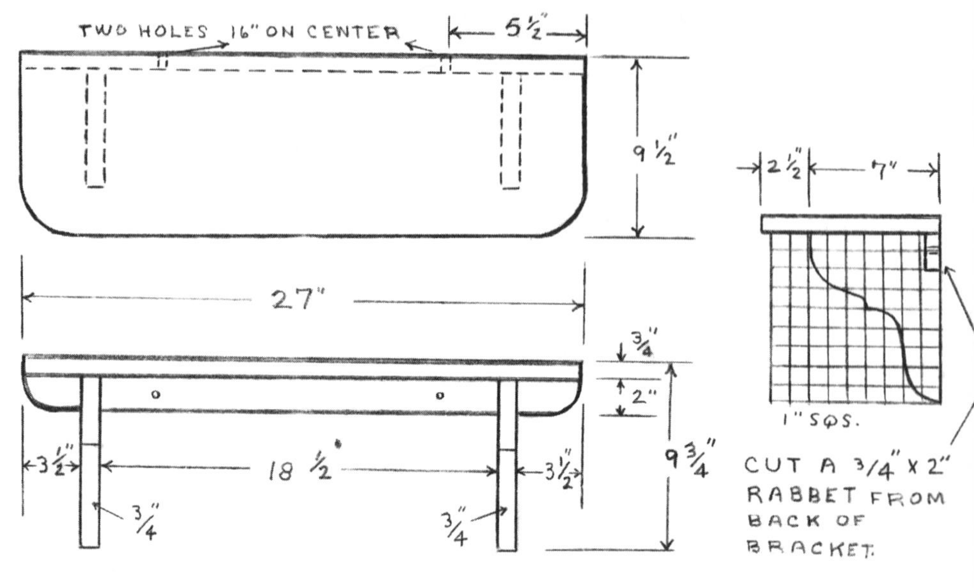

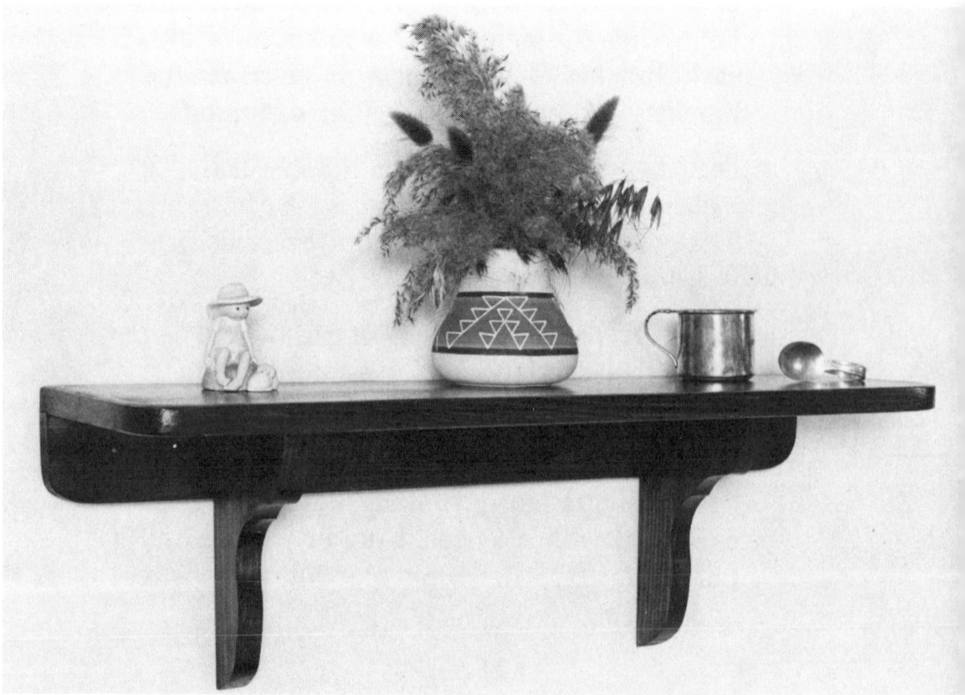

## Project 25

# END TABLE

### Materials Required

*Wood of Your Choice*

1 **top** ¾″ × 18″ × 27½″ oval
2 **legs** ¾″ × 8″ × 19¼″
2 **bases** 1¼″ × 3″ × 18″
2 **upper cleats** ¾″ × 1⅜″ × 11″
1 **shelf** ¾″ × 6″ × 12″
2 **shelf cleats** ¾″ × 1⅜″ × 4½″
2 **decorative side rails to shelf** ¾″ × 1½″ × 12″
4 **wood screws** #14, 3″ flathead
22 **wood screws** #10, 1¼″ flathead

### Procedures

1. Glue the ¾-inch stock together for the required widths for all pieces, then make patterns from the 1-inch squares for all parts.

2. Lay out and cut the leg assembly to the design suggested. This includes the base, the leg, and the upper cleat. Round all edges to a ¼-inch-round cut using a router and ¼-inch-round bit or by hand with a spokeshave and half-round file.

3. Secure the base to the leg by drilling two counterbored pilot screw holes up through the base and into the leg. Fasten with #14, 3-inch flathead wood screws.

4. Lay out and cut all pieces of the shelf assembly to the suggested size and design.

# End Table

CLEAT
3/4" X 1 1/2" CLEAT SUPPORT.
18"
11"
3 1/2"
1" SQ
TOP PATTERN
1" SQ.
23"
27 1/2"
1 3/8"
3/4
3"
12"
11 1/2
1 1/4"
1/2 ROUND ALL EDEGES
OPTIONAL STRAIGHT DESIGN

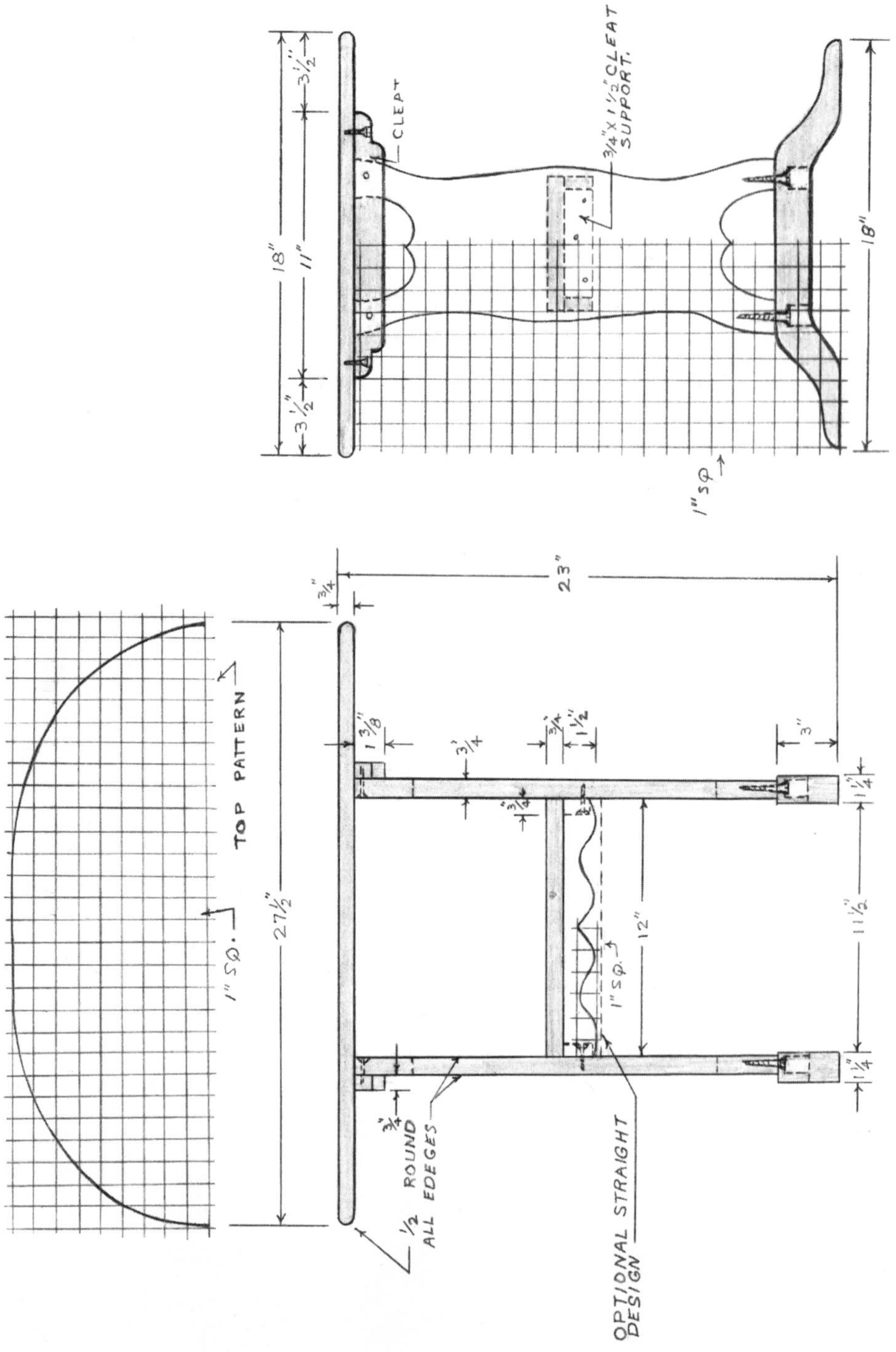

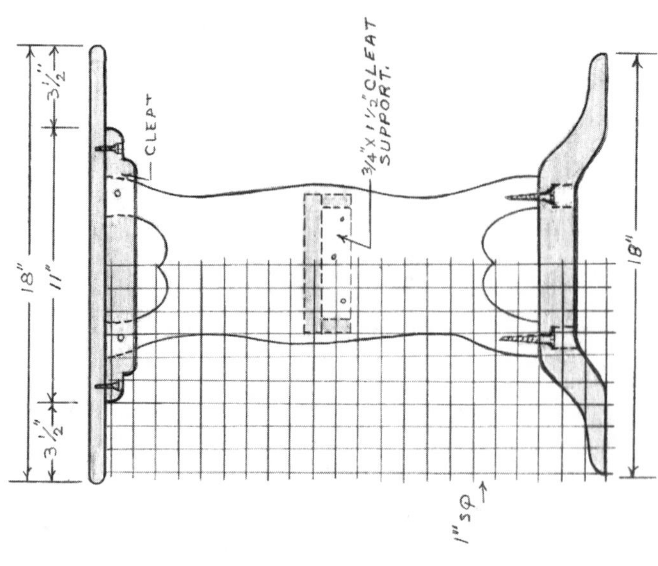

Fasten the shelf assembly together by gluing and clamping the decorative side rail and end cleats to the bottom of the shelf. For added reinforcement, drill counterbored pilot holes through the underside of the rail pieces and cleats, then fasten #10, 1¼-inch flathead wood screws to the shelf.

5. Lay out and cut the oval top to the recommended size.

   File and sand the edge smooth. Continue to ¼-inch round-cut the edge using a router and ¼-inch round bit, or by hand using a spokeshave and file.

6. Secure the shelf unit to the legs by drilling three 3/16-inch-diameter holes through each cleat.

   Position the shelf assembly onto one leg and fasten with #10, 1¼-inch flathead wood screws.

   Fasten the other leg to the shelf unit by holding it in its proper location using two barclamps, then attach three #10, 1¼-inch flathead wood screws.

7. To complete the total assembly, fasten the oval top to the cleats attached to the upper legs. Drill two counterbored pilot screw holes through each cleat and slightly into the top, then secure the top with #10, 1¼-inch flathead wood screws.

8. Thoroughly sand the entire table, then select and apply the colored stain of your choice. When completely dry, apply four coats of finishing oil, rubbing lightly between dry coats with fine steel wool or pumice and oil.

## Project 26

# FLOOR LAMP WITH TABLE

## Materials Required

*Wood of Your Choice*

**1 lamp post** 2″ × 2″ × 47″
**1 table** ¾″ × 16″ × 16″
**1 top to base** ¾″ × 10″ × 10″
**1 bottom to base** ¾″ × 14″ × 14″
**4 post braces** ¾″ × 3½″ × 6″
**8 wood screws** #10, 1½″ flathead
**1 wood screw** #12, 3″ flathead
**12 furniture buttons** ⅜″ dia.

*Electrical Hardware:*
**lamp cord** 8′ to 10′, 120 volt
**threaded connecting pipe** ⅜″ dia. × 1½″
**on/off electric socket with shade**

*Note:* All of the hardware listed above can be purchased in an easy-to-assemble kit.

## Procedures

1. Use bar clamps to edge-glue enough rough stock to meet the required width for the table and base pieces.

   Cut all pieces to the size and shape shown in the drawing.

2. To make the slot for the cord in the pole, cut a ⅜-inch groove the length of the pole toward the back of the two boards before gluing them up to form the 2-inch by 2-inch width.

   *Note:* A hole should be drilled through the base for the cord and another groove cut into base so the cord can run out to the receptacle.

# Floor Lamp With Table

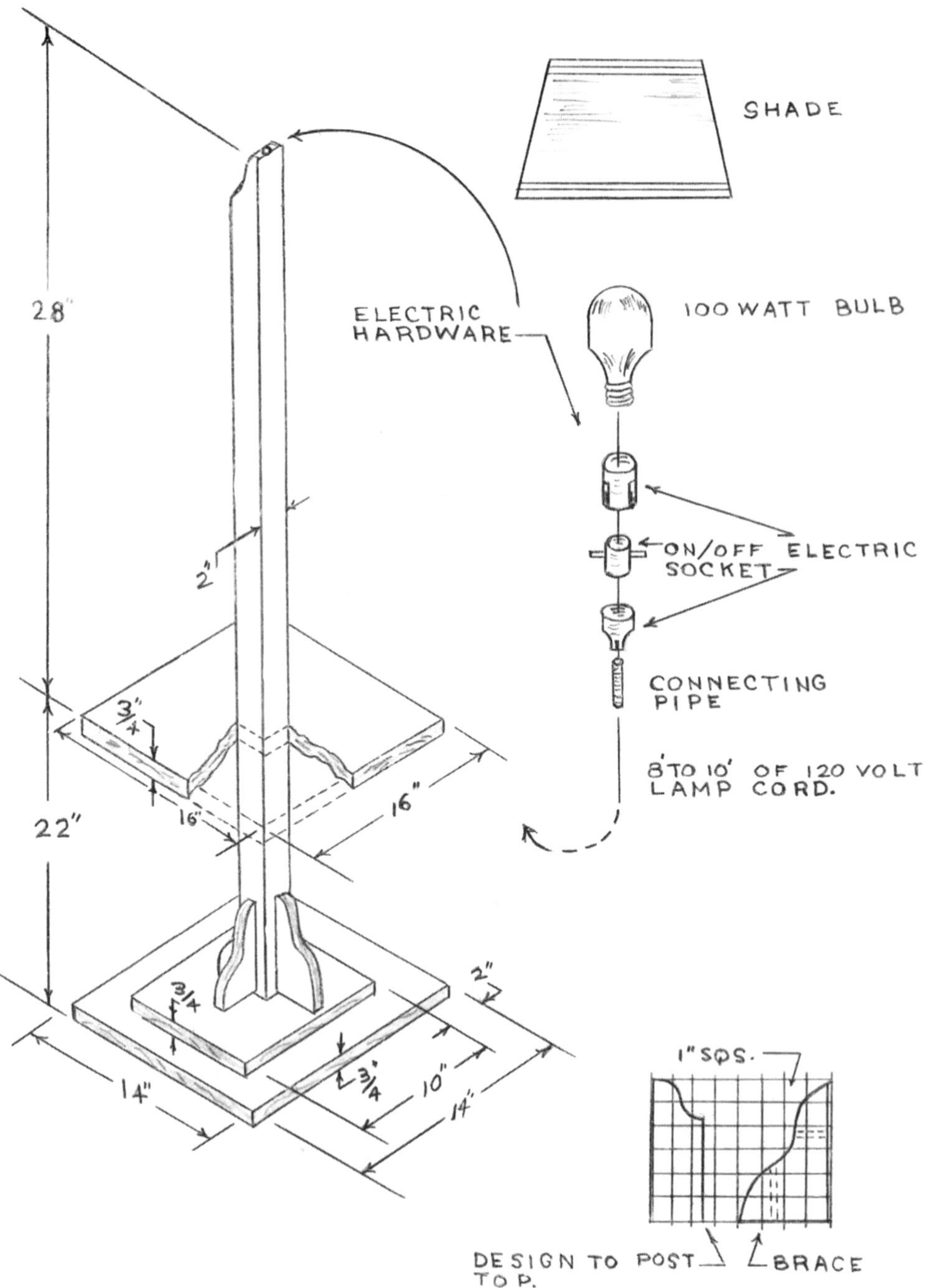

3. On the base and table you may want an edge design. A chamfer cut, cove cut, or rounded edge is attractive on this project. These cuts can be made by hand with a jack plane and file, or they can be made with a power router.
4. Fasten the base together by gluing and clamping the top to the bottom.

   Drill a countersunk pilot hole up through the center of the base and into the bottom of the pole, then proceed to attach the pole to the base with a #12, 3-inch flathead wood screw. Be sure that the ⅜-inch by ⅜-inch groove cut through the post is matched up to the ⅜-inch bored hole in the base to allow the cord to pass through.
5. Drill two ⅜-inch counterbored pilot screw holes into each brace, then fasten the four braces to the pole for support. One hole should be drilled into the pole and the other drilled into the base.
6. Lay out and cut a 2½-inch by 2½-inch square hole in the center of the table.

   Secure the table approximately 18 inches to 22 inches above the floor. Use 1-inch glue blocks fastened to the pole beneath the table, then glue and nail the table to the 1-inch glue blocks.
7. Install the 1½-inch by ⅜-inch-diameter threaded connecting pipe into the precut opening at the top of the pole. If necessary, lock the pipe in place by applying epoxy glue to the bottom section of the threaded pipe.

   Fasten the bottom of the electric socket to the threaded pipe.

   Run the lamp cord through the precut ⅜-inch hole in the center post.

Cut back the wire to approximately ¾ inch, then tie the wires to the terminals for the on/off switch.

*Note:* To prevent the cord from falling through, tie a knot in the cord within the bottom part of the three-part socket. Snap the three-part electric socket closed. Attach an alternating current plug on the bottom end of the cord.

8. Install ⅜-inch-diameter dowel plugs or furniture buttons into the exposed counterbored holes with glue.
9. Sand the lamp to a smooth finish. Apply the stain or paint of your choice. When dry, apply several coats of clear polyurethane finish, rubbing between coats with fine steel wool. Preserve the finish with an occasional coat of paste wax or lemon oil.
10. Finish the assembly by installing the lamp shade. The shade size should not be larger than the base or it will look awkward.

## Project 27

# PLANT STAND

### Materials Required

*Wood of Your Choice*

**1 base** ¾″ × 14″ dia.
**4 bases** ½″ × 3″ dia.
**4 plates** ½″ × 2½″ dia.
**4 tops** ½″ × 6″ dia.
**1 dowel** ½″ × 12″
**1 dowel** ½″ × 18″
**1 dowel** ½″ × 24″
**1 dowel** ½″ × 30″

### Procedures

1. Lay out and cut all of the pieces to the suggested diameters.
2. Locate and fasten the four ½-inch by 2½-inch-diameter plates beneath the four stand tops.
3. Locate and fasten the four ½-inch by 3-inch-diameter bases to the large 14-inch-diameter base.
4. Lay out and bore a ½-inch hole to a depth of ¾ inch through the center of the four 2½-inch-diameter underplates and the four 3-inch-diameter bases.
5. Before attaching the ½-inch dowels, sand all of the pieces to a smooth surface.
6. Insert glue into the prebored holes, then assemble the dowel to the plates and base.

# Plant Stand

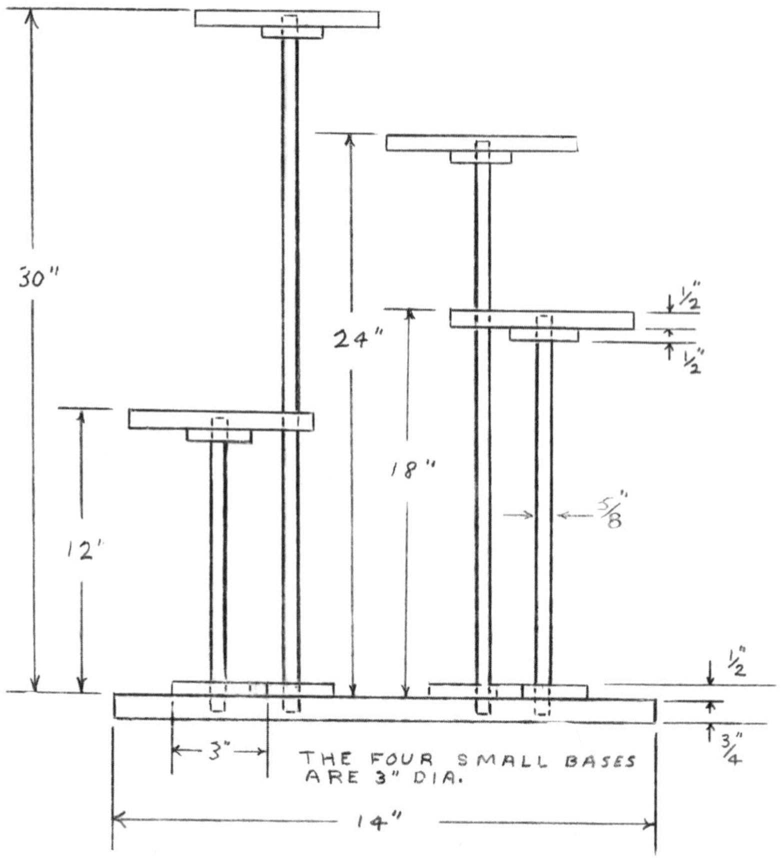


7. Select and apply the stain or paint of your choice. When dry, apply several coats of clear polyurethane or lacquer finish, rubbing between each coat with fine steel wool or pumice and oil. Protect the finish with a coat of paste wax and buff.

## Project 28

# CARPENTER'S TOOL BOX

### Materials Required

*Wood of Your Choice*

**2 ends** ¾″ × 8″ × 10″
**2 sides** ½″ × 7½″ × 31½″
**1 bottom** ½″ × 7½″ × 31½″
**2 tops** ½″ × 5″ × 31⅞″
**1 handle** 1¼″ dia. × 33″
**2 outside handle supports** ½″ × 4″ × 15″
**2 continuous hinges** 1″ × 30½″

### Procedures

1. Lay out and rough cut all of the stock to size. If necessary, edge-glue the stock to reach the required width.
2. Lay out and cut the two ¾-inch by 8-inch by 10-inch ends to the top middle point shown in the drawing. If using a power saw, set the miter to 32 degrees. Cut ¼-inch by ½-inch rabbets on the two edges of each end piece.
3. Cut the two ½-inch by 7½-inch by 31-inch side pieces. Bevel the top edge to a 32-degree angle to match the angle of the end pieces.
4. Cut a ¼-inch by ½-inch rabbet around the inner bottom edge of the two ends and sides to receive the bottom.
5. Plane, scrape, and sand the five pieces to the box assembly, then fasten them together by gluing and clamping. If necessary, reinforce by drilling counterbored pilot screw holes then inserting #8, 1½-inch, flathead wood screws.

# Carpenter's Tool Box

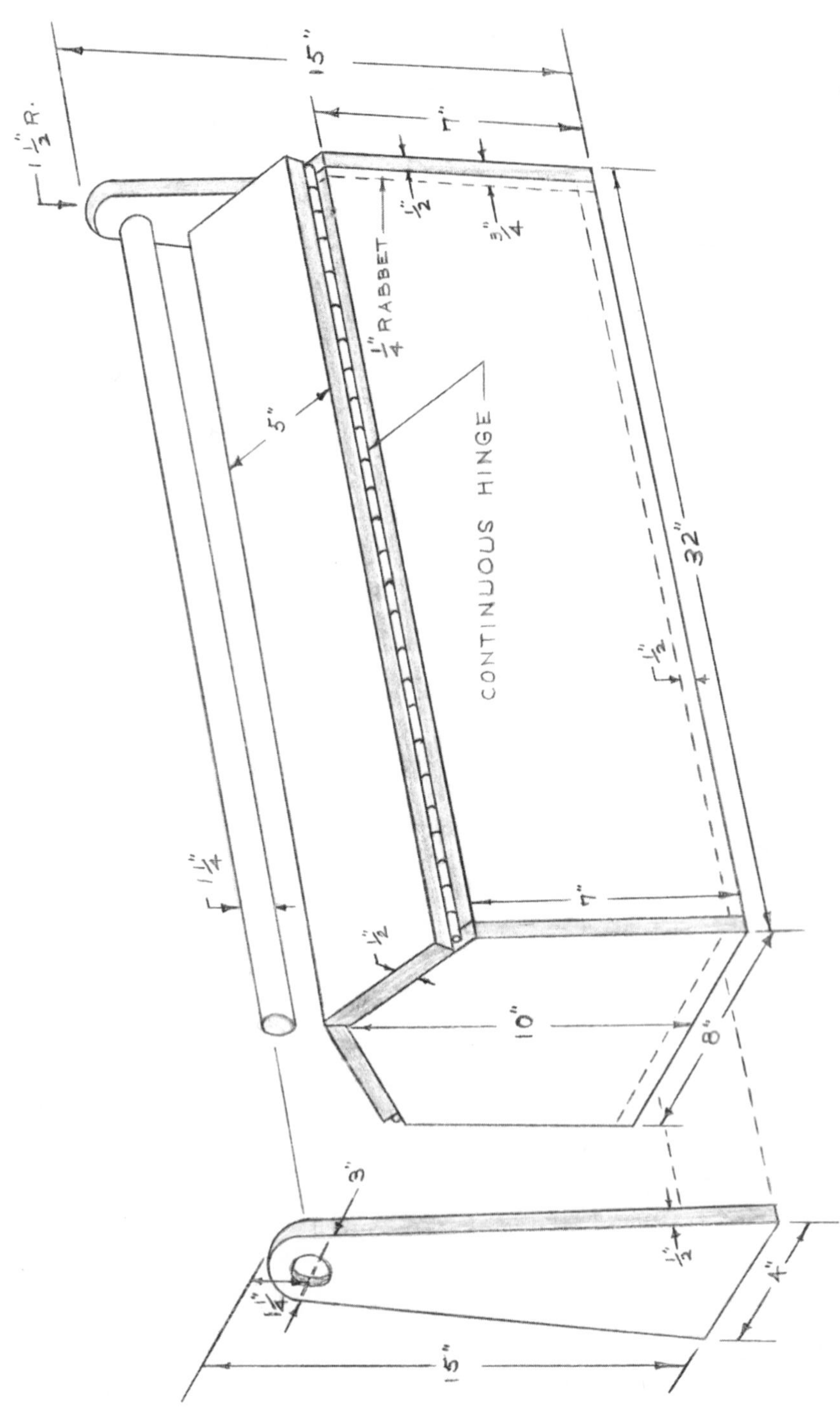

Cover holes with furniture plugs. Wipe off all traces of glue with a wet cloth.

6. Lay out and cut the two ½-inch by 5-inch by 31⅞-inch tops. Use a plane or saw to cut a bevel at the top edge of each piece. Secure the tops to the sides by installing the 1-inch by 31½-inch continuous hinges.
7. Lay out and cut the two outside handle supports to the size and design indicated in the drawing. Taper from 4 inches down to 3 inches, then bore a 1¼-inch-diameter hole approximately 1¼ inches down from the top edge of the supports to receive the handle.
8. Insert the 1¼-inch-diameter by 33-inch dowel handle into the supports, then fasten the supports to the ends by drilling counterbored pilot screw holes and installing #10, 1-inch flathead screws. Plug holes with furniture plugs.
9. Either purchase or make the handles for the two tops. If you construct the handles, start by squaring up two 2¼-inch by 3-inch by 5-inch walnut blocks. Bore a 1¼-inch-diameter hole completely through the block from end to end. The hole should be located a little off-center. Locate and cut the block so that the saw kerf is made into the far side of the 1¼-inch bored hole. Gape the handle by filing and sanding. Secure the handles to the two tops by drilling counterbored pilot holes then installing #8, ¾-inch flathead screws. Cover holes with ⅜-inch birch furniture plugs.
10. Fill all holes, scrape off any traces of glue, then proceed to sand the tool box to a final smooth surface. Leave the wood natural or

stain to your color choice. Apply four coats of clear oil finish, using fine steel wool or #400 emery paper to rub between coats. On the final coat rub the finish down with lemon oil and pumice. Preserve the finish with an occasional coat of lemon oil or paste wax.

## Project 29

# SPICE CABINET

### Materials Required

*Wood of Your Choice*

**2 sides** ¾″ × 4½″ × 29¼″
**1 top** ½″ × 4½″ × 12″
**1 bottom** ½″ × 4½″ × 12″
**3 shelves** ½″ × 4¼″ × 10½″
**3 dowels** ⅜″ dia. × 12″
**1 back** ¼″ × 11¼″ × 29¼″

*Large Inner Frame:*
**1 top** ¾″ × 2½″ × 12″
**1 bottom** ¾″ × 2½″ × 12″
**2 sides** ¾″ × 2½″ × 24¼″

*Small Outer Frame:*
**1 top** ½″ × 1¼″ × 11″
**1 bottom** ½″ × 1¼″ × 11″
**2 sides** ½″ × 1¼″ × 23¼″
**1 crown design piece** ½″ × 4″ × 11″
**1 molding above crown** ¾″ × 1¾″ × 12″
**overall drawer size** 5″ × 4″ × 10½″ with a ½″ overlap on the front panel

**2 butt hinges**
**2 door knobs**
**1 plate glass** ⅛″

### Procedures

1. Lay out and cut all pieces for the cabinet frame and shelves.

# Spice Cabinet

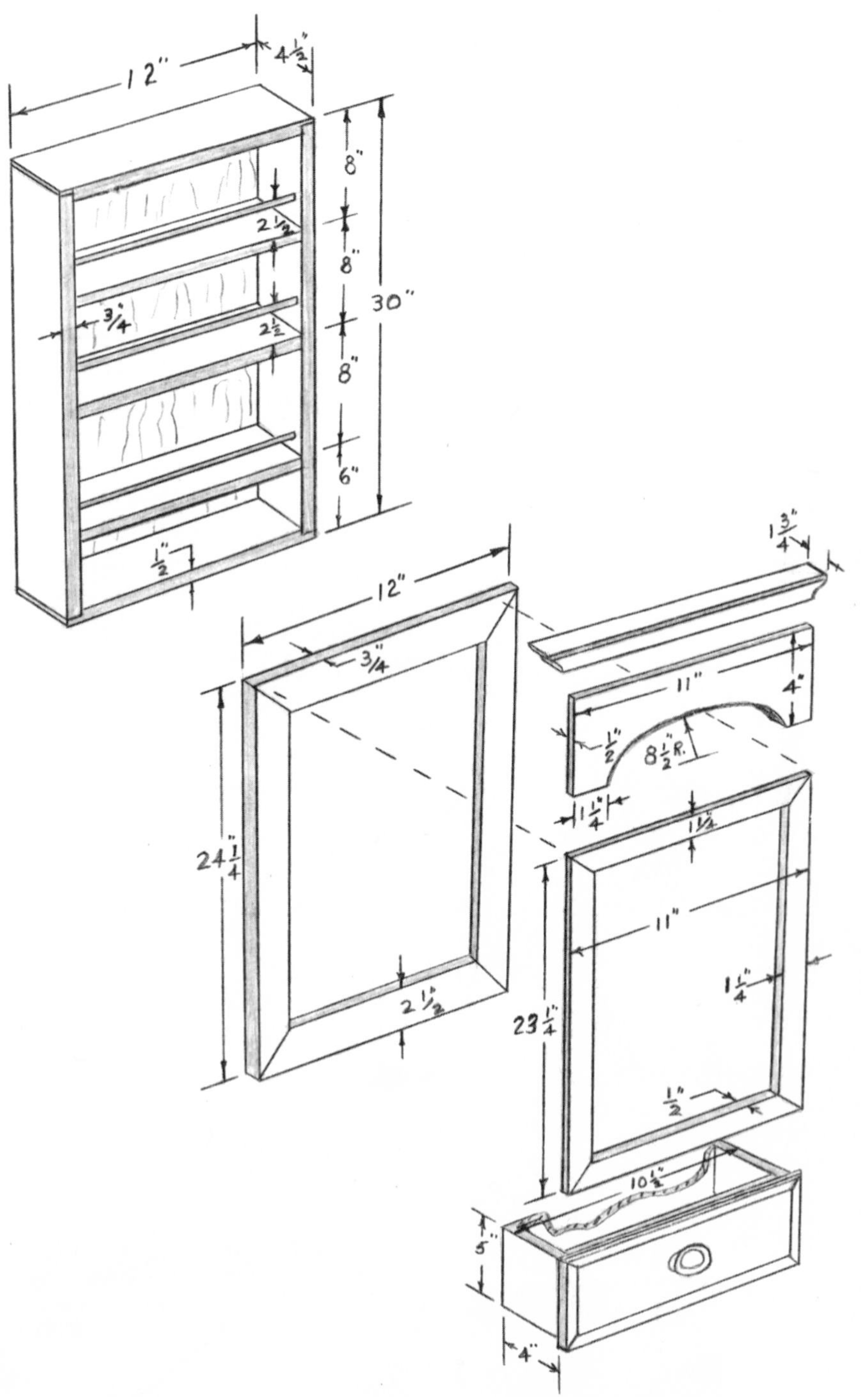

Cut a ¼-inch by ⅜-inch rabbet on the rear inner edge of the four frame pieces to receive the back.

Cut ¼-inch by ¾-inch rabbets at the ends of the top and bottom to receive the side pieces.

Lay out and drill completely through the sides the matching holes for the ⅜-inch-diameter dowels.

Lay out the location for the shelves, then continue to fasten the assembly together using glue and #4 finishing nails. Use ⅞-inch brads to secure the back. Use glue to secure the dowels.

2. Lay out and cut all pieces for the large and small frame.

   Cut a ⅜-inch by ⅜-inch rabbet on the inner edges of the four pieces on the large frame to receive the glass.

   Cut 45-degree angled miters for the frame corner joints.

   Fasten the two frames together using dowels to reinforce the miter joints.

   For added design use a router to make a cove or chamfered cut around the outside edge of each frame.

   Secure the small frame to the larger frame by gluing and clamping. Wipe off excess glue with a wet cloth.

3. Lay out and cut the crown design on the top piece, then proceed to cut a cove design around the three exposed edges of the top molding piece. The exposed edges of the crown pieces are rounded over to a quarter-round design.

4. Construct the small drawer to the overall size given in the drawing. The front should overlap ½ inch all around. Refer to the chapter on drawer construction for helpful suggestions.
5. Hang the door by installing two 1½-inch butt hinges.
6. Set and fill all holes, then sand the cabinet to a smooth finished surface. Select and apply the stain or paint of your choice. When dry, apply three coats of polyurethane finish, rubbing between coats with 320 emery paper or fine steel wool. Preserve the finish with lemon oil.
7. Install the door and drawer pulls of your choice. Install ⅛-inch glass with glass points or back it with ¼-inch by ⅜-inch strips of wood.

## Project 30

# TOWEL RACK CABINET

### Materials Required

*Wood of Your Choice*

**1 back** ¾″ × 11½″ × 29½″
**2 sides** ¾″ × 8″ × 16″
**1 top shelf** ¾″ × 7¼″ × 28″
**1 bottom shelf** ¾″ × 7¼″ × 28″
**1 drawer divider** ¾″ × 4″ × 7¼″
**1 dowel** ½″ × 29½″
**2 knobs** ¾″ dia.

*Drawer:*
**2 fronts** ¾″ × 3¹⁵⁄₁₆″ × 13½″
**4 sides** ⅜″ × 3¹⁵⁄₁₆″ × 7″
**2 backs** ⅜″ × 3¹⁵⁄₁₆″ × 12¾″
**2 bottoms** ⅜″ × 6¼″ × 12¾″
**finishing nails** #6
**finishing brads** 1″

### Procedures

1. Lay out and cut all pieces to the suggested size and design.
2. Bore two ½-inch holes into the side pieces then insert the ½-inch dowel with glue.
3. Fasten the top shelf and bottom piece to the sides and back using #6 finishing nails. Locate the center, then attach the drawer divider by gluing and clamping.
4. Lay out and cut the ¼-inch by ⅜-inch rabbet joints on the two drawer front pieces, then proceed to assemble the drawers to the size suggested, using 1-inch finishing brads.

# Towel Rack Cabinet

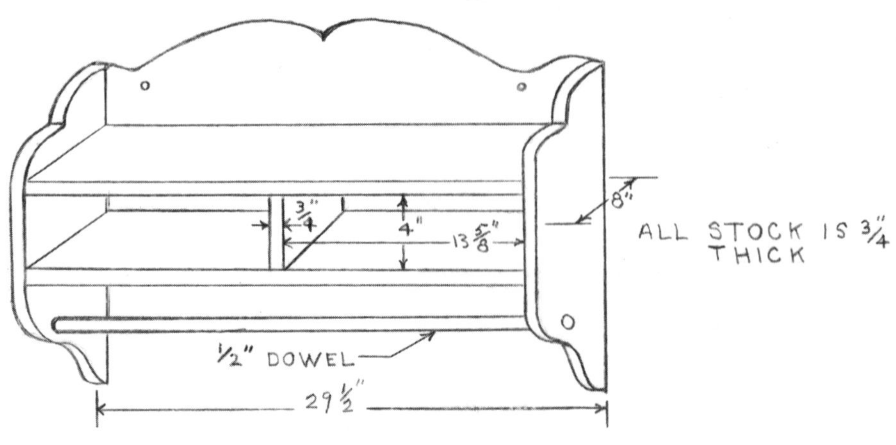


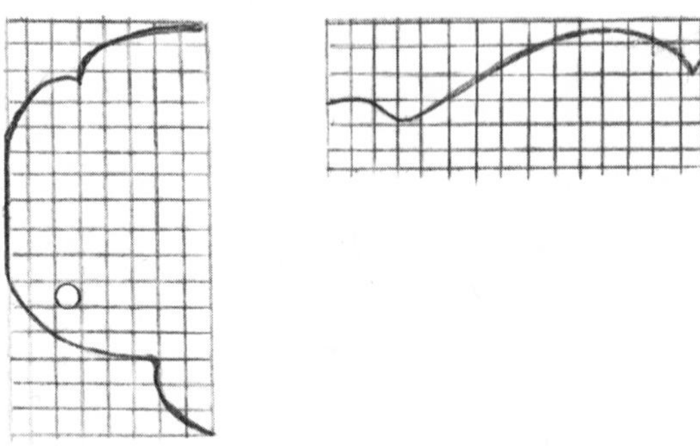


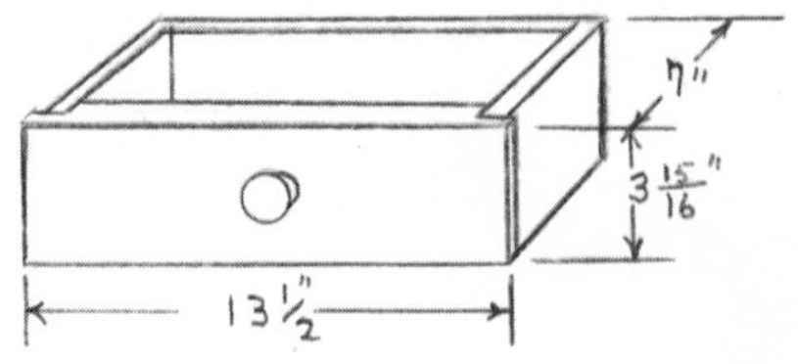

5. Attach the two pull knobs to the center of each drawer.
6. Drill two $\frac{3}{16}$-inch holes 16 inches apart to hang the towel rack cabinet to the wall.
7. Set and fill all nail holes, then sand the project to a smooth surface.
8. Select and apply the stain or paint of your choice. When dry, apply several coats of clear polyurethane finish. Rub between each coat with fine steel wool or pumice and oil. Protect the finish by applying a coat of paste wax and buffing.

## Project 31

# WALL SCONCE

### Materials Required

*Wood of Your Choice*

**1 back** ¾″ × 5″ × 20″
**1 bracket** ¾″ × 4″ × 4½″
**1 top disk** ¾″ × 2³⁄₁₆″ dia.
**1 bottom disk** ¾″ × 5″ dia.
**1 candle cup** ¾″ dia.
**3 wood screws** #9, 1½″ flathead
**1 globe** 2¼″ dia. base

### Procedures

1. Lay out and cut the four pieces to the shape and size shown in the drawing.
2. Bore a ¾-inch-diameter hole through the center of the top disk. If a ¾-inch-diameter candle cup is to be used, bore a ⅞-inch-diameter hole through the top disk.
3. Glue and clamp the top disk to the center of the bottom disk.

   Drill a pilot hole through the center of the bottom disk and into the bracket.

   Secure the two glued disks to the bracket with a #9, 1½-inch flathead screw.
4. Fasten the bracket and disk assembly to the back piece by drilling two countersunk pilot screw holes through the back and into the bracket. Use #9, 1½-inch flathead screws.
5. Round over all edges with a file, then sand the sconce to a smooth finished surface.

# Wall Sconce

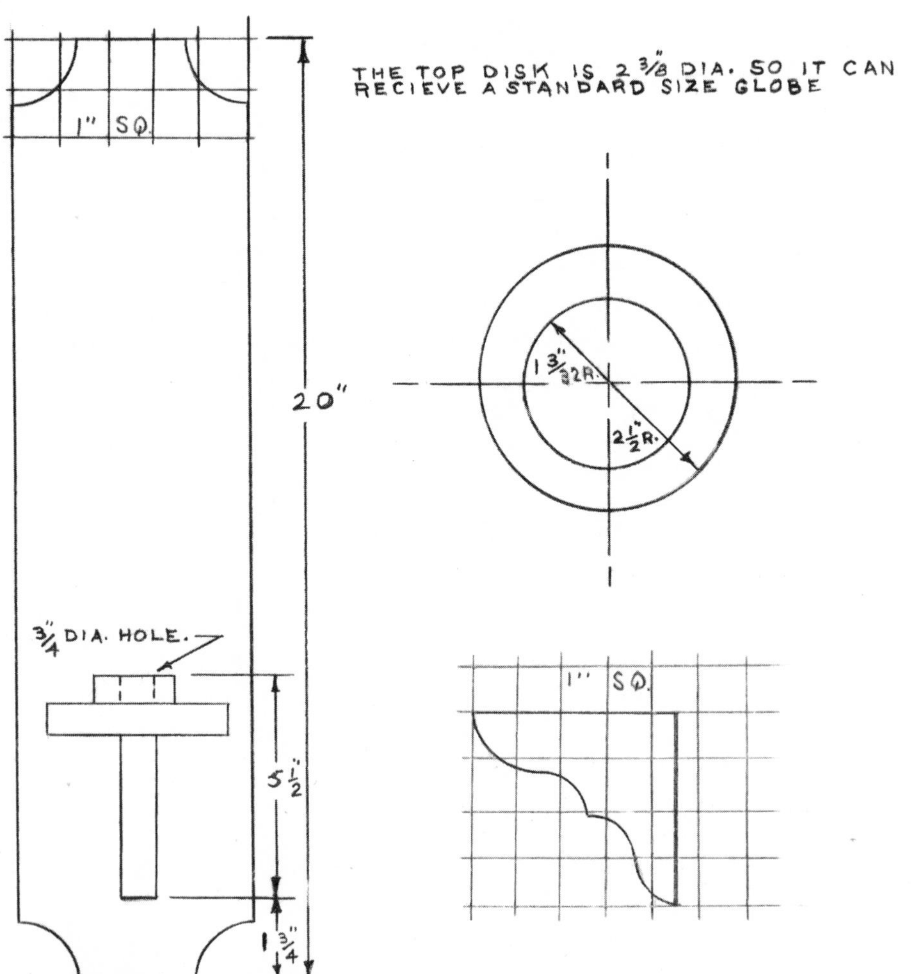


6. Select and apply the stain or paint of your choice, then cover with three coats of lacquer, rubbing the final coat down with pumice and lemon oil. Preserve the finish with occasional coats of lemon oil.

## Project 32

# BUD VASE

### Materials Required

*Wood of Your Choice*

**1 vase block** 2″ × 2″ × 12″
**1 test tube** ⅝″ dia. or ¾″ dia.

*Note:* A test tube is only necessary if live flowers are to be used. It should be mentioned that this attractive project can be made in a multitude of shapes and sizes by simply varying the cutout section from a radius to a straight design or to the shape of your choice.

### Procedures

1. Cut a piece of stock to the thickness, width, and length of your choice.
2. Bore the diameter hole of your choice to a depth that will exceed the cutout section by approximately 1 inch.

   *Note:* If a test tube is to be used, bore the hole 1/16 inch larger than the diameter of the tube.
3. With a hand saw, sabre saw, or coping saw, cut out the shape of your choice.
4. File and sand the bud vase to a smooth finished surface. A small sanding drum attached to an electric drill is helpful in sanding a radius.
5. Apply four coats of clear finishing oil, rubbing between each coat with pumice and lemon oil. Preserve the finish with an occasional coat of paste wax or lemon oil.

   *Note:* If a test tube is not used, a ¾-inch-diameter brass candle cup will add beauty to the vase.

# Bud Vase

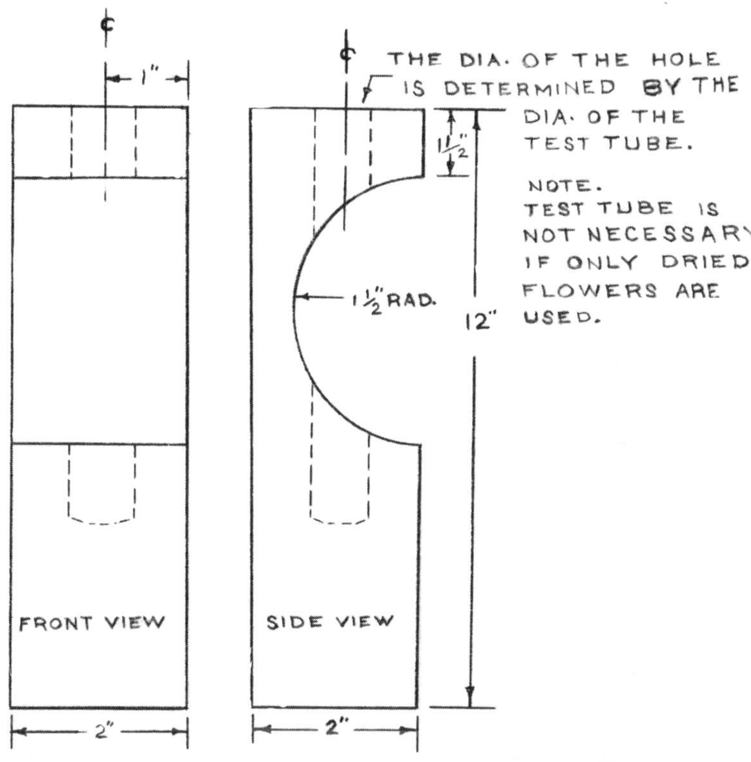
¢
1"
¢
THE DIA. OF THE HOLE IS DETERMINED BY THE DIA. OF THE TEST TUBE.
1½"
NOTE.
TEST TUBE IS NOT NECESSARY IF ONLY DRIED FLOWERS ARE USED.
1½" RAD.
12"
FRONT VIEW
SIDE VIEW
2"
2"

## Project 33

# CORNER SHELF

### Materials Required

*Wood of Your Choice*

**2 sides** ½″ × 6″ × 17½″
**1 top shelf** ½″ × 3″ × 6″
**1 middle shelf** ½″ × 3¾″ × 7½″
**1 bottom shelf** ½″ × 4¾″ × 9″
**finishing nails #4**

### Procedures

1. Lay out and cut the two side pieces to ½ inch by 6 inches by 17½ inches. Set the saw to a 45-degree angle and cut the miter joint on the side edge for the corner joint.
2. Lay out and cut the design on the sides and shelves.
3. Fasten the two sides by gluing and nailing into the miter joint with #4 finishing nails.
4. Attach the three shelves to the sides by gluing and nailing with #4 finishing nails.
5. Drill two 3⁄16-inch-diameter holes through the sides to hang the corner shelf.
6. Scrape off all traces of glue. Set and fill all nail holes, then sand the corner shelf to a smooth surface.
7. Select and apply the colored stain or paint of your choice. When dry, apply several coats of lacquer, rubbing between each coat with fine steel wool or pumice and oil. To protect the finish apply a coat of paste wax and buff.

# Corner Shelf

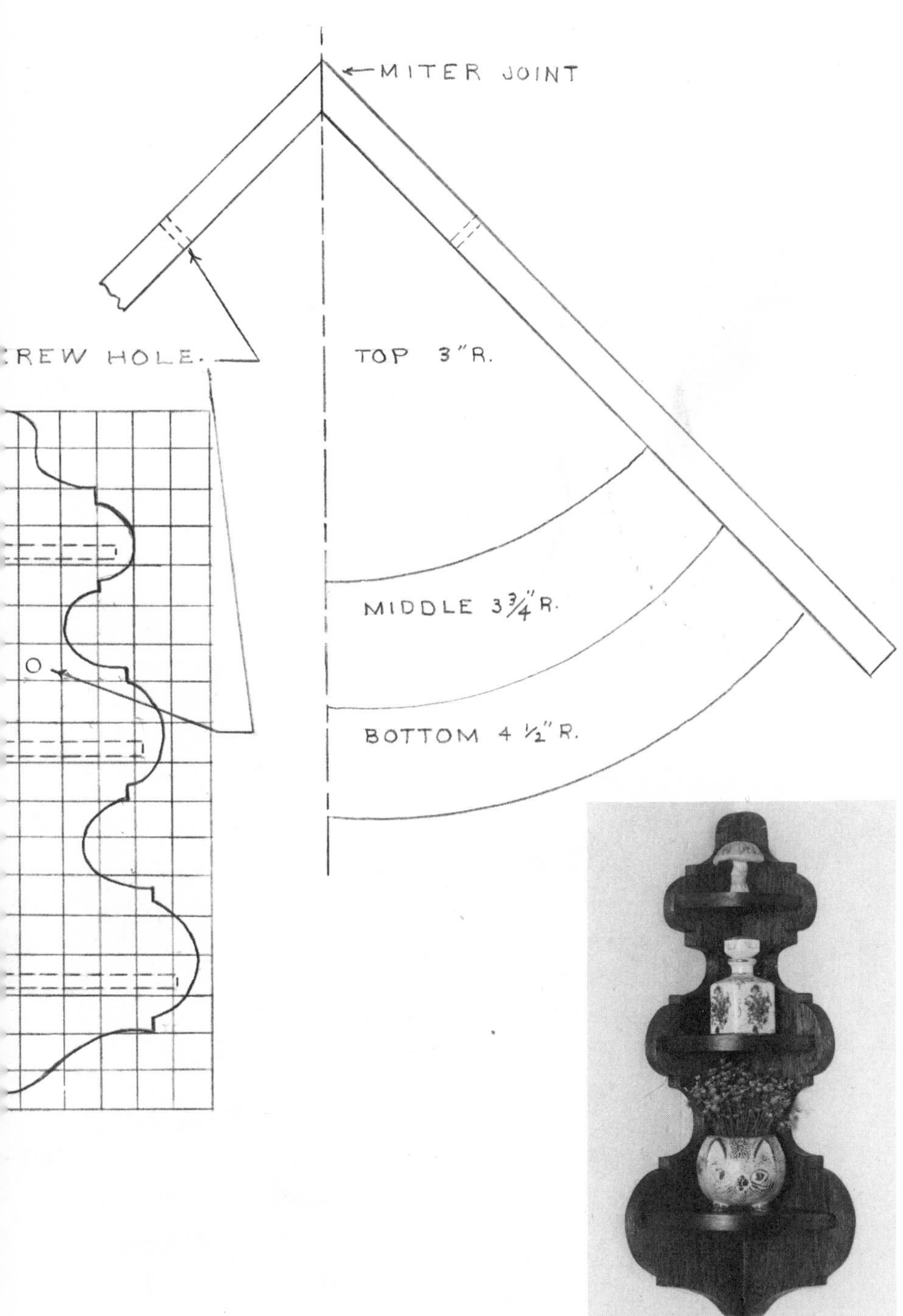

## Project 34

# CHILD'S CRADLE

### Materials Required

*Wood of Your Choice*

**2 sides** ¾″ × 21″ × 36″
**1 bottom** ¾″ × 12″ × 29½″
**1 back** ¾″ × 18″ × 21″
**1 front** ¾″ × 14″ × 11″
**1 front to top section** ¾″ × 5″ × 18″
**1 top middle piece** ½″ × 9″ × 12½″
**2 top slanting pieces** ½″ × 6½″ × 12½″
**2 rockers** ¾″ × 6″ × 20″
**nails** #6 finishing or #6 antique cut wrought iron
**wood screws** #12, 1½″ flathead

### Procedures

1. Rough cut the stock, then edge-glue enough stock to meet the required widths for all pieces necessary.

2. Cut the two side pieces to the size and design shown in the drawing. The top edge of the sides should be cut to approximately 37 degrees. The bottom edge of the sides should be cut to approximately 10 degrees.

3. Cut the bottom out to the size indicated in the drawing. All four sides should be cut to a 10-degree angle.

4. Fasten the two sides to the bottom by drilling counterbored pilot holes and screwing with #12, 1½-inch screws or by using #6 nails. Cover holes with hardwood plugs.

# Child's Cradle

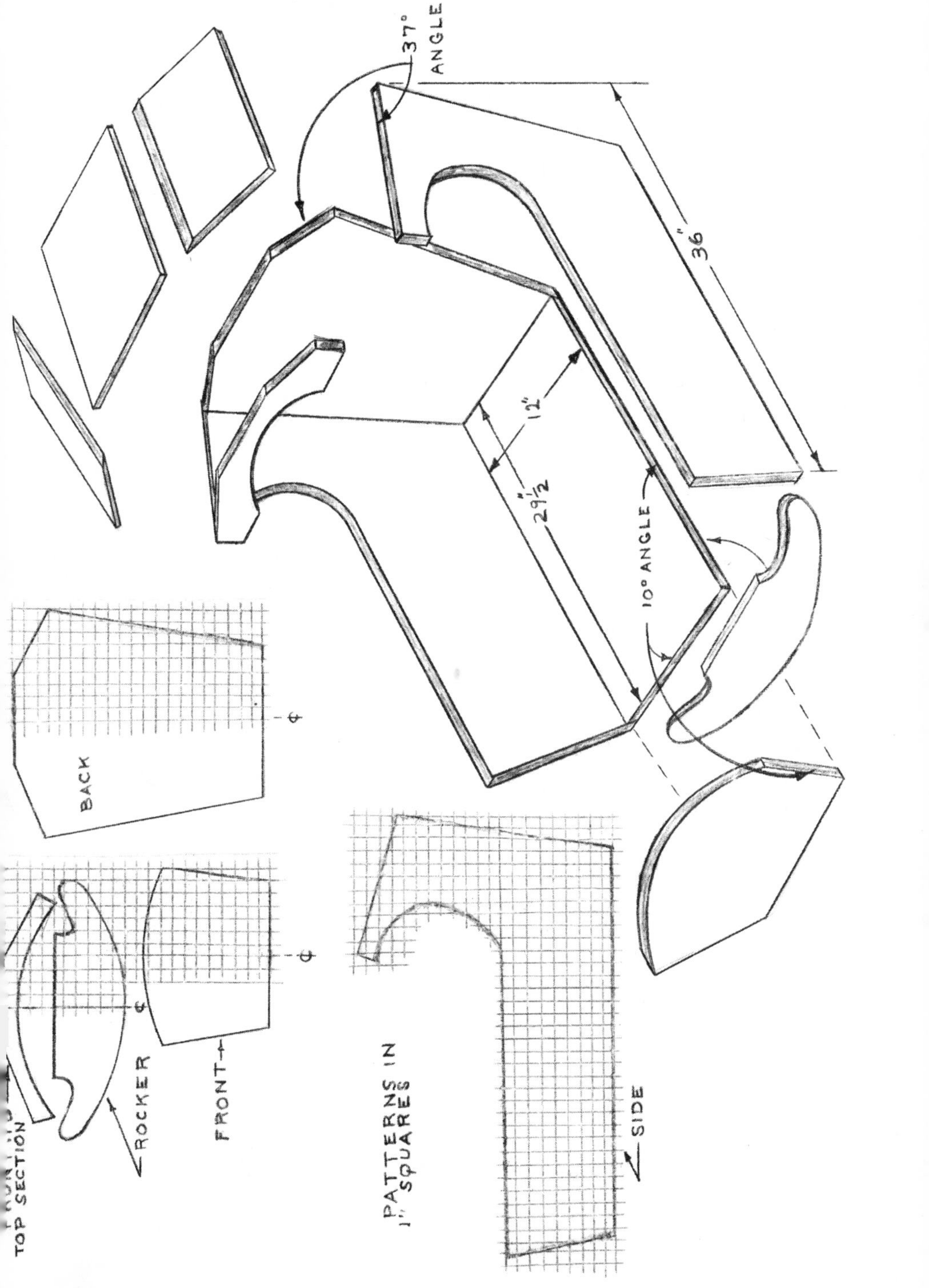

5. Lay out and cut the back, front, and the front of the top section to the pattern shape shown in the drawing. Make sure the angles on the front of the top section match the angles to the back for the three top pieces to fit properly. The side edges on the front and back must taper down at approximately 10 degrees.

   *Note:* Before cutting the patterns, it may be wise to match them up to the preassembled sides and bottom for checking the correct 10-degree angle.

   Fasten these three pieces to the assembly using counterbored screws or nails.

6. Cut the three top pieces to correct size. To allow the top to overhang the sides, you will have to plane the top edge of the sides to match the angle of the middle top section. Secure in place with nails or screws in counterbored pilot holes.

7. Cut the two rockers to the size and shape indicated in the drawing. Fasten them to the bottom with #12, 1½-inch flathead screws recessed into counterbored pilot holes.

8. Cover all screw heads with hardwood plugs. Set and fill nail holes. If antique nails are used, leave them exposed, then sand the project to a smooth surface.

   Select and apply the stain or paint of your choice. When dry, apply three coats of clear polyurethane finish, rubbing between coats with #400 emery paper or fine steel wool. On the final coat rub down with lemon oil and pumice. Preserve the finish with an occasional coat of paste wax or lemon oil.

## Project 35

# CHEST/END TABLE

### Materials Required

*Wood of Your Choice*

**1 top** 1″ × 18″ × 30″
**1 front** ¾″ × 8″ × 30″
**1 back** ¾″ × 8″ × 30″
**2 ends** ¾″ × 8″ × 17¼″
**1 bottom** ½″ × 16½″ × 28½″
**1 drawer runner** ¾″ × 1″ × 17″
**4 legs** 2″ × 4″ × 6½″

*Optional Base:*
**1 front** ¾″ × 6″ × 31½″
**1 back** ¾″ × 6″ × 31½″
**2 ends** ¾″ × 6″ × 19½″

**1 drawer** (overall size) 5⅞″ × 16″ × 23⅞″ with a ⅜″ lip around the front panel

### Procedures

1. Start by using bar clamps to edge-glue up enough stock for the necessary pieces.
2. Lay out and cut the front, back, and two ends to the size and shape shown in the drawing.

   Cut ⅜-inch by ¾-inch rabbets at the ends of the front and back pieces.

   Fasten the four pieces together using glue and #6 finishing nails through the rabbet joints.
3. Lay out and cut the bottom to the exact dimensions given in the list of materials.

# Chest/End Table

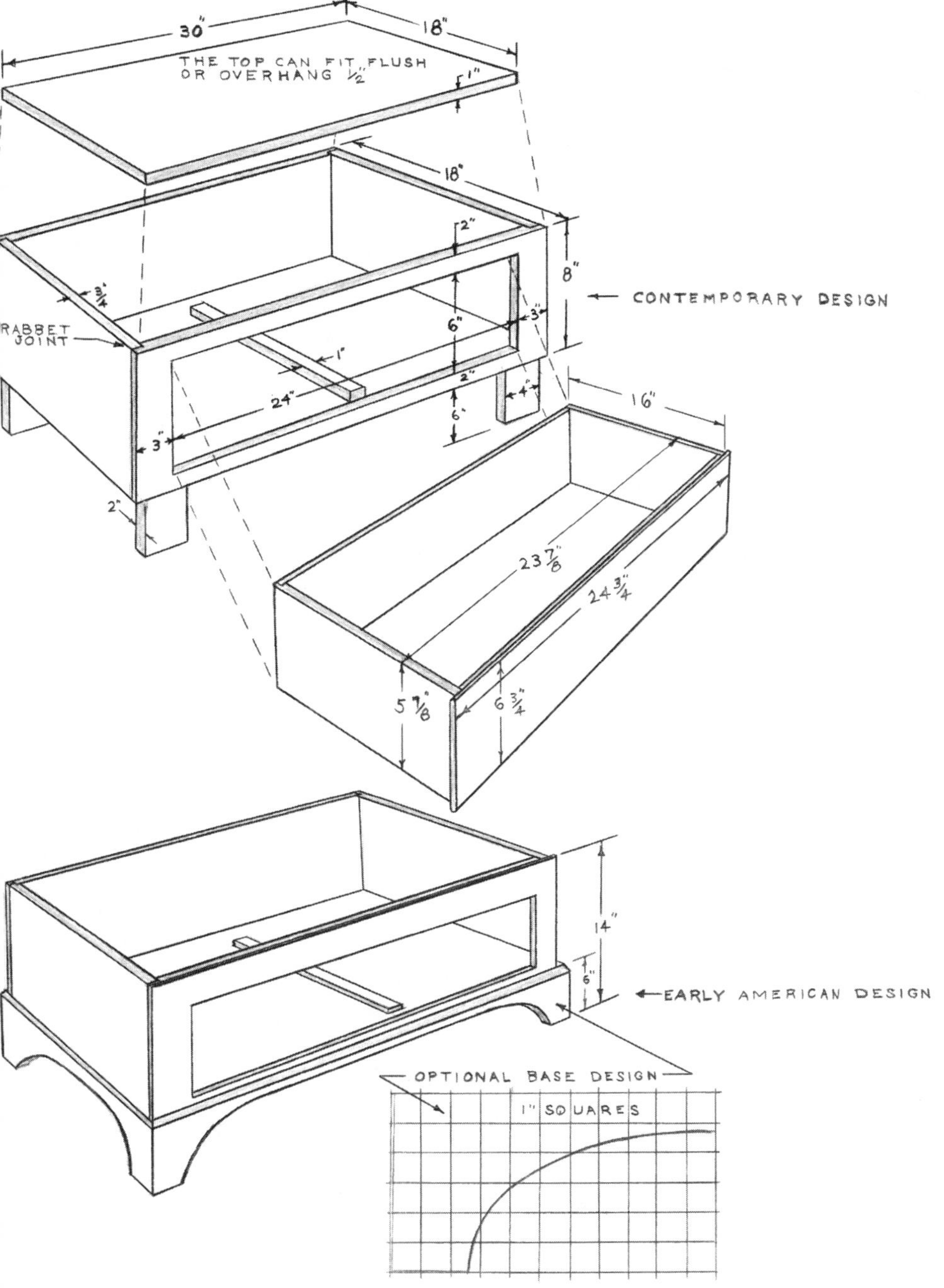

Fasten it to the inner frame of the four pieces using glue and #6 finishing nails. The top edge of the bottom should rest flush to the top edge of the drawer opening.

*Note:* With a wet cloth wipe off all traces of glue.

4. Construct the ¾-inch by 1-inch by 17-inch drawer runner and fasten it to the center.
5. Cut the four legs to size, then secure them to the inner corners beneath the bottom.

   Drill pilot holes, then fasten with #10, 2½-inch flathead screws.

   If the optional base design is used, cut out the shape, make a bevel on the top edge, then cut the corners at a 45-degree angle. Secure into place with #6 finishing nails and glue.

6. Construct the drawer to the overall size given in the drawing. The front should have a ⅜-inch overlap all around. Refer to the front of the book for instructions on drawer assembly.
7. Set and fill all nail holes, scrape off all traces of glue, then sand the table to smooth final finish.
8. Select and apply the stain or paint of your choice. When dry, apply four coats of clear finishing oil, rubbing between coats with #320 emery paper. On the final coat rub down with pumice and lemon oil. Preserve the finish with an occasional coat of lemon oil.
9. Select and install the hardware of your choice.

## Project 36

# STEP STOOL

### Materials Required

*Wood of Your Choice*

**1 top** 3/4" × 8 1/2" × 16"
**2 sides** 3/4" × 13 1/2" × 10 3/4"
**1 step shelf** 3/4" × 12 1/2" × 14 1/4"
**1 front** 3/4" × 5 1/2" × 13 1/2"
**1 back** 3/4" × 2" × 13 1/2"
**nails** #6 finishing
**wood screws** #10, 1 1/2" flathead
**dowel plugs or furniture buttons** 3/8"

### Procedures

1. Lay out and cut all pieces to the suggested size and shape.
2. Lay out and cut the two 3/8-inch by 3/4-inch dado joints, then fasten the step shelf to the side pieces with glue and #6 finishing nails.
3. Fasten the front and back pieces to the sides, then locate and drill four 3/8-inch countersunk pilot holes through the top and into the side pieces. Attach the top to the assembly with #10, 1 1/2-inch flathead wood screws. Fill the countersunk holes with 3/8-inch dowel plugs or 3/8-inch furniture buttons.
4. Scrape off all traces of glue. Set and fill all nail holes, then sand the stool to a smooth surface.
5. Select and apply the stain or paint of your choice. When dry, apply several coats with fine steel wool or pumice and oil. Protect the finish by applying a coat of paste wax and buffing.

# Step Stool

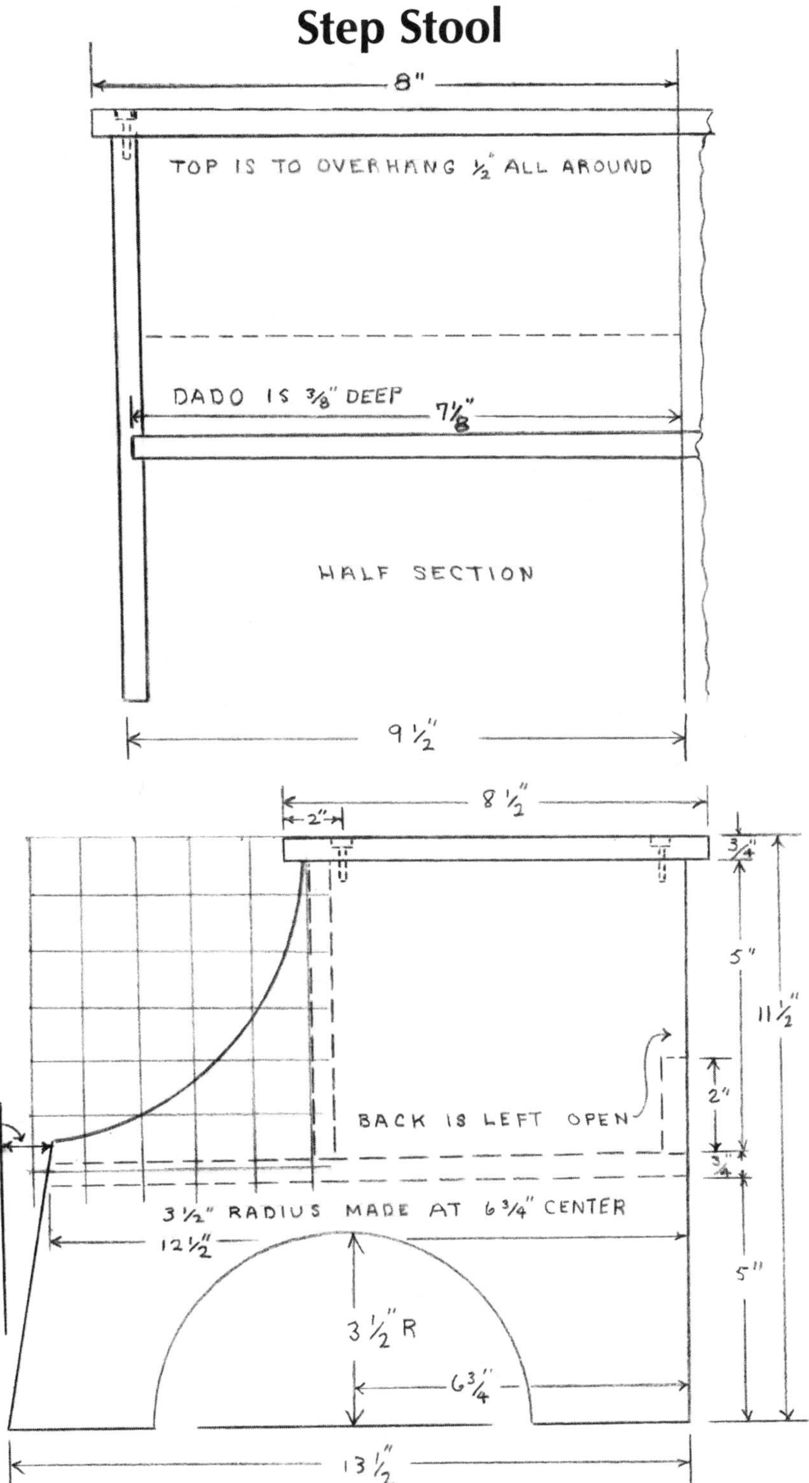
8"
TOP IS TO OVERHANG ½" ALL AROUND
DADO IS ⅜" DEEP
7⅛"
HALF SECTION
9½"
8½"
2"
¾"
5"
11½"
2"
BACK IS LEFT OPEN
¾"
¾"
3½" RADIUS MADE AT 6¾" CENTER
12½"
5"
3½" R
6¾"
13½"

## Project 37

# SPOON RACK/PLANTER

### Materials Required

*Wood of Your Choice*

**1 back** ½″ × 13″ × 28″
**2 sides** ½″ × 4″ × 5¾″
**1 front** ½″ × 4″ × 13″
**1 bottom** ½″ × 5″ × 12″
**2 spoon brackets** ¾″ × 1¼″ × 13″
**finishing nails #4**

### Procedures

1. Glue up the stock to the required width, then lay out and cut all of the pieces to the suggested size and design.
2. Lay out and cut the two ¼-inch by ½-inch rabbet joints on the front of the box. Fasten the sides and bottom to the front with glue and #4 finishing nails. Fasten the box assembly to the back by gluing and clamping.
3. Attach the spoon brackets to the back by gluing and clamping. The spoon slots can be cut into a rectangular shape as shown in the drawing, or a ⅜-inch-diameter hole can be drilled to receive the spoon.
4. Drill a ¼-inch hole to hang the spoon rack.
5. Scrape off all traces of glue. Set and fill all nail holes, then sand to a smooth surface.
6. Select and apply the stain or paint of your choice. When dry, apply several coats of clear polyurethane finish, rubbing between each coat with fine steel wool or pumice and oil. To preserve the finish apply a coat of paste wax and buff.

# Spoon Rack/Planter

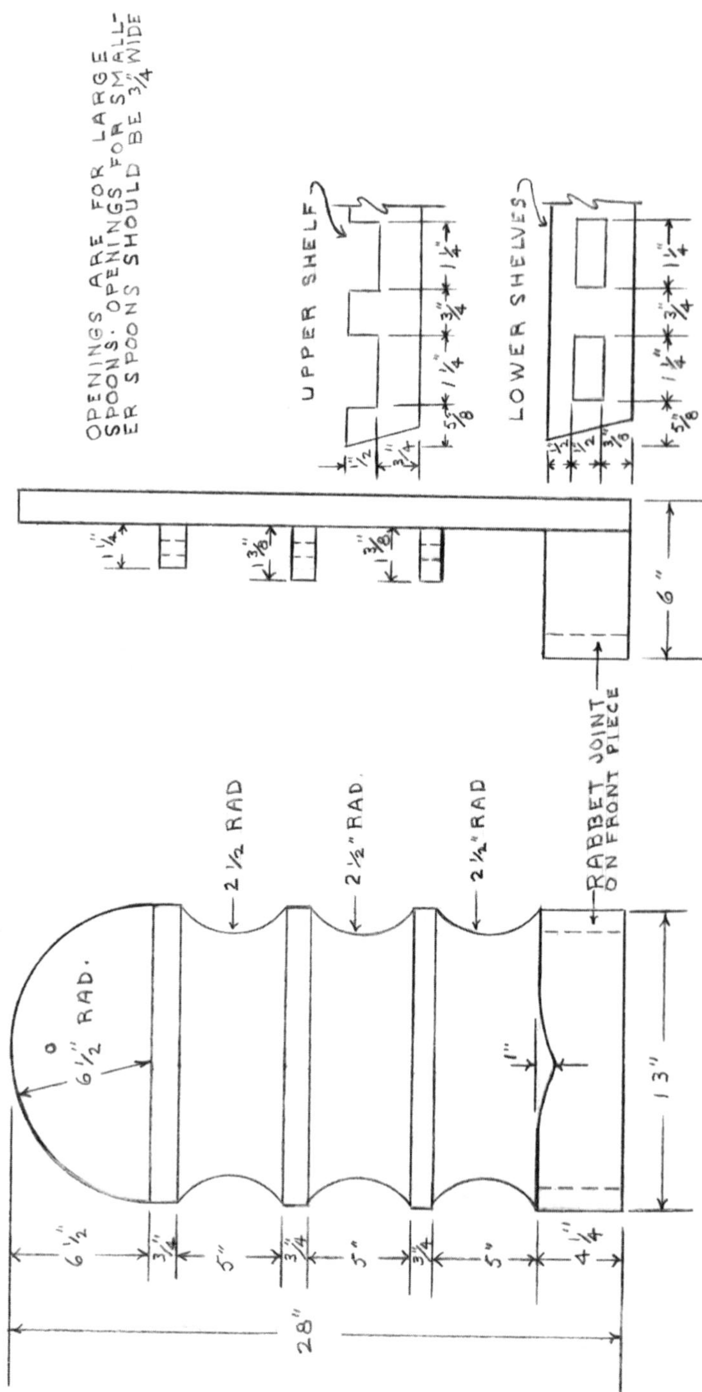

## Project 38

# KITCHEN CABINETS

### Materials Required

(Number of pieces required depends on the number of cabinets to be made.)

*Wood of Your Choice*

*Lower Cabinets:*

**upper rails** ¾″ × 2″ × length as needed
**bottom rails** ¾″ × 4″ × length as needed
**frame stiles** ¾″ × 2″ × 35″
**divider panels** ¾″ × 22½″ × 35″
**end panels** ¾″ × 23¼″ × 35″
**shelves** ¾″ × 23¼″ × length as needed
**countertop** 1″ × 26″ × length as needed
**splash board** ¾″ × 5″ × length as needed
**doors** ¾″ × 19⅞″ × 33″
**drawer runners** ¾″ × 2″ × 23″
**drawer parts** dimensions determined by the number of doors and their size

*Note:* The exposed doors, end panels and drawer panels can be made from ¾″ birch plywood or glued-up stock of your choice. The rails and stiles can be solid birch or wood of your choice. The inner divider panels and shelves can be ¾″ particle board or fir plywood.

*Upper Cabinets:*

**rails** ¾″ × 2″ × length as needed
**frame stiles** ¾″ × 2″ × 28″
**divider panels** ¾″ × 12¼″ × length as needed
**end panels** ¾″ × 12¼″ × 28″
**doors** ¾″ × 19⅞″ × 28″

*Note:* The doors are figured on a 20″ frame opening.

**hardware** 2½″ surface or butt hinges at two per door, with a magnetic catch for each door.

# Kitchen Cabinets

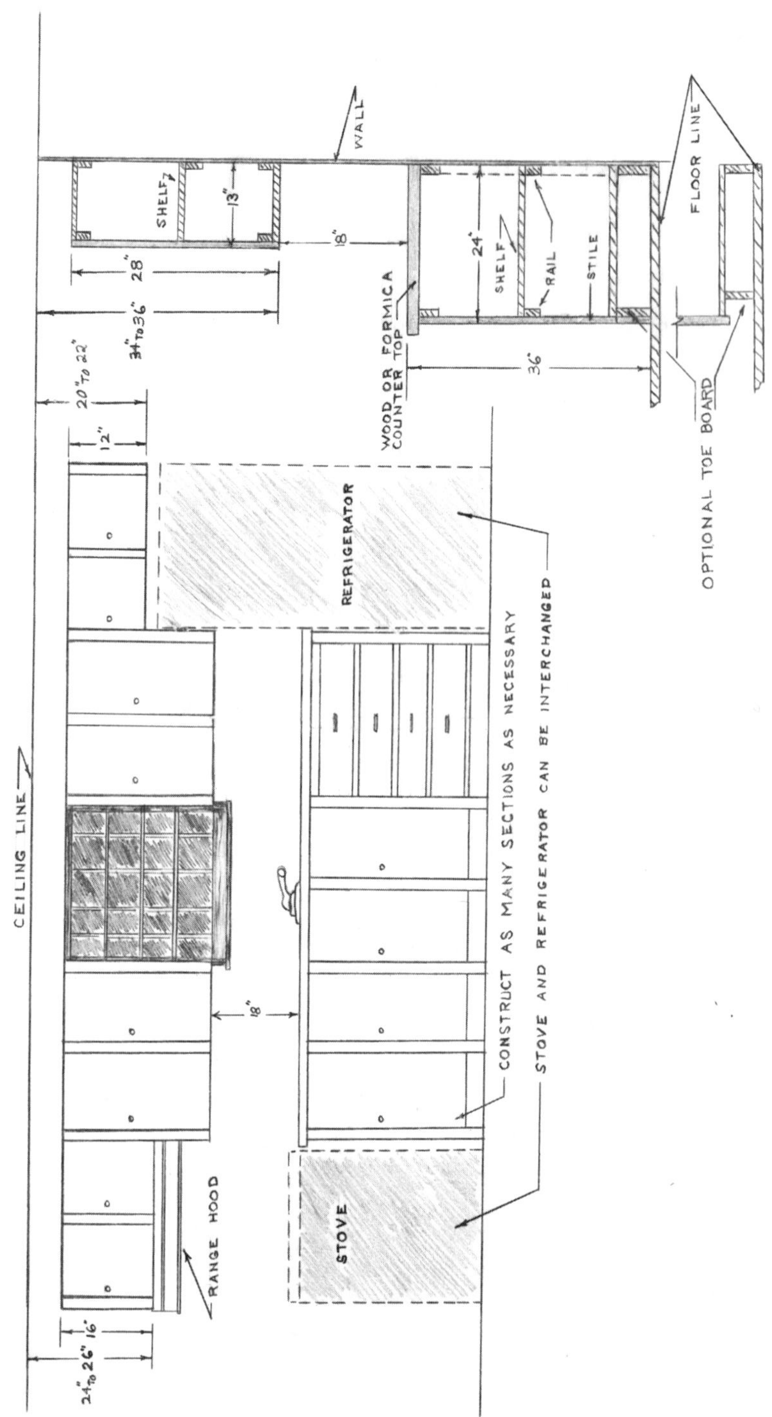

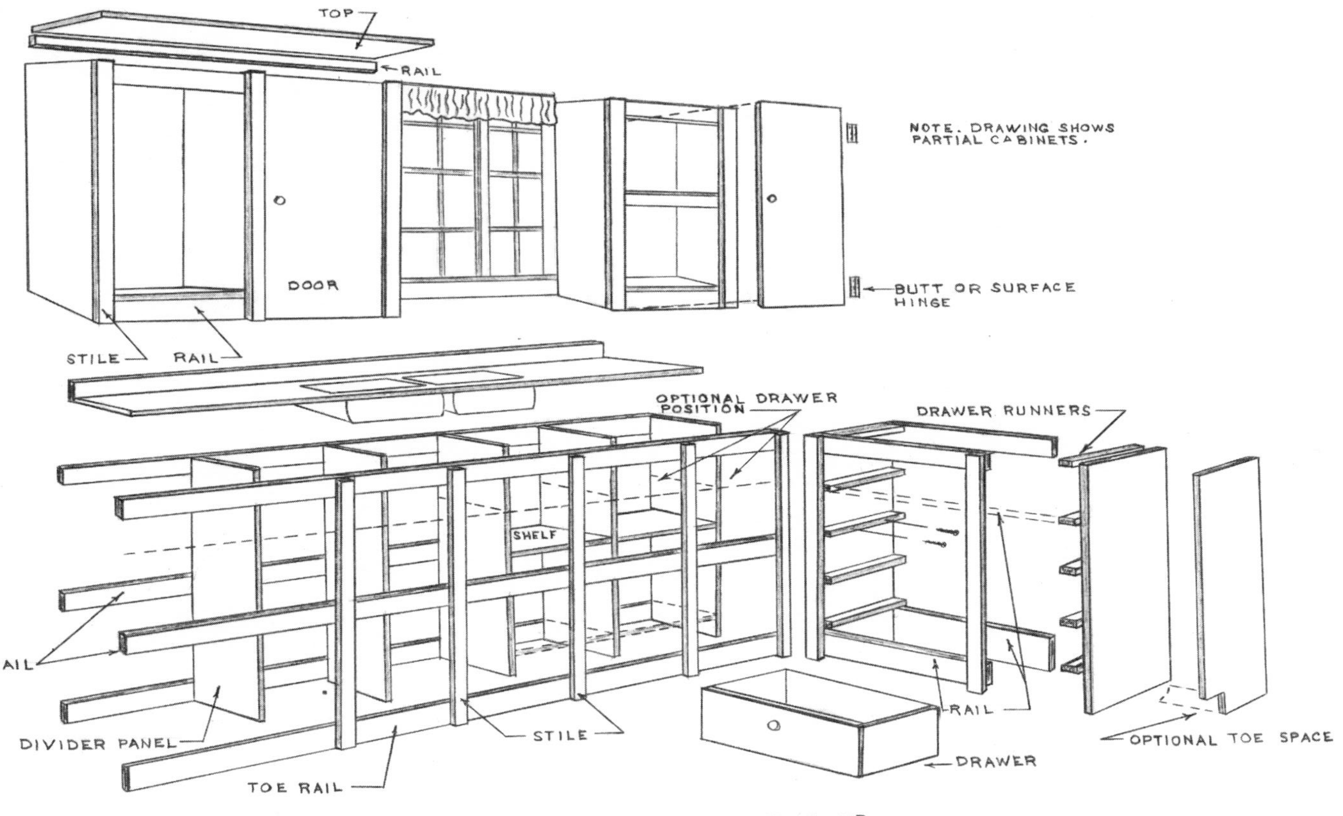
TOP
RAIL
NOTE. DRAWING SHOWS PARTIAL CABINETS.
DOOR
BUTT OR SURFACE HINGE
STILE
RAIL
OPTIONAL DRAWER POSITION
DRAWER RUNNERS
SHELF
RAIL
DIVIDER PANEL
STILE
RAIL
OPTIONAL TOE SPACE
DRAWER
TOE RAIL
THESE CABINETS CAN BE PREFABRICATED OR CONSTRUCTED DIRECTLY TO THE WALL

## Procedures

1. Begin by cutting all of the rails for the lower cabinets to the length determined by your kitchen size.

   Secure the rear rails to the rear edge of the end/dividing panels. The top edge of the rails should be located at 18 inches and 35 inches above the floor.

   Fasten the assembly to the wall by driving #8 nails through the rails and into the wall studs. Rails should be screwed to the dividing panels with #10, 1½-inch flathead screws.

   Fasten the three front rails to the front edge of the dividing panels at the same spacing as the back rails. For reinforcement fasten the bottom toe rail to a ¾-inch by 2-inch cleat fastened to the floor.

2. Lay out and cut the stiles. Fasten them to the rails at a 20-inch spacing, using glue and #6 finishing nails. Make sure the stiles are perpendicular to the rails (90 degrees).

3. Lay out and cut the shelves to be placed between the divider panels. Mount them to the top edge of the rails by gluing and nailing with #6 finishing nails.

4. Lay out and cut the countertop to size. Use contact cement to secure the plastic laminate.

   Lay out and make the cutout for the sink using a sabre saw with a fine tooth blade. Cut the splash board to size, then cover it with plastic laminate. Fasten it to counter by using #10, 3-inch flathead screws driven up from the bottom of the counter.

5. Locate position for hinges, then hang the doors. Secure the door pulls and magnetic catches into place.

6. Construct the drawer cabinets using the stiles and rails method used to make the shelf cabinet.

   Use the drawer dividers for spacing four to five drawers. If the drawers are to remain flush with the front of the cabinet, use ¾-inch by 1-inch spacer rails mounted to the front edge of the runners. Allow approximately ⅛ inch of clearance between drawer fronts and spacer rails. Refer to section in book on drawer construction.

   In constructing the upper cabinets follow the same directions as for the lower cabinets. The rails and stiles are all ¾ inch by 2 inches.

   *Note:* Countertops and splash boards can be purchased premade with sink cutouts from your local lumber yard.

7. Set and fill all holes, then sand the cabinets smooth. If the cabinets are constructed of fir, they can be painted. If constructed of birch, use stain or a clear finish.

## Project 39

# KNIFE HOLDER

### Materials Required

*Wood of Your Choice*

**1 base** ½″ × 3″ × 7¾″
**1 front section** 2″ × 6¾″ × 8½″
**1 back section** ½″ × 6¾″ × 8½″
**4 screws** #10, 1¼″ flathead

### Procedures

1. Lay out and cut the three pieces to the suggested size. You may have to glue up stock to achieve the correct thickness.
2. Lay out the exact location for the knife slots. Cut the slots on the back side of the front piece, making sure that the saw blade cut is to the correct depth for each knife slot. Fasten the back piece to the front to enclose the knife slots. Use glue and clamps.
3. Lay out and cut a ¼-inch by ½-inch chamfer or cove design around the top edge of the base.
4. Fasten the base to the bottom of the knife holder by drilling pilot and countersunk holes and screwing four #10, 1¼-inch flathead screws.
5. Clean off all traces of glue and proceed to sand the project to a smooth surface.
6. Apply several coats of clear finishing oil, rubbing between each coat with fine steel wool or pumice and oil. Protect the finish with paste wax.

# Knife Holder

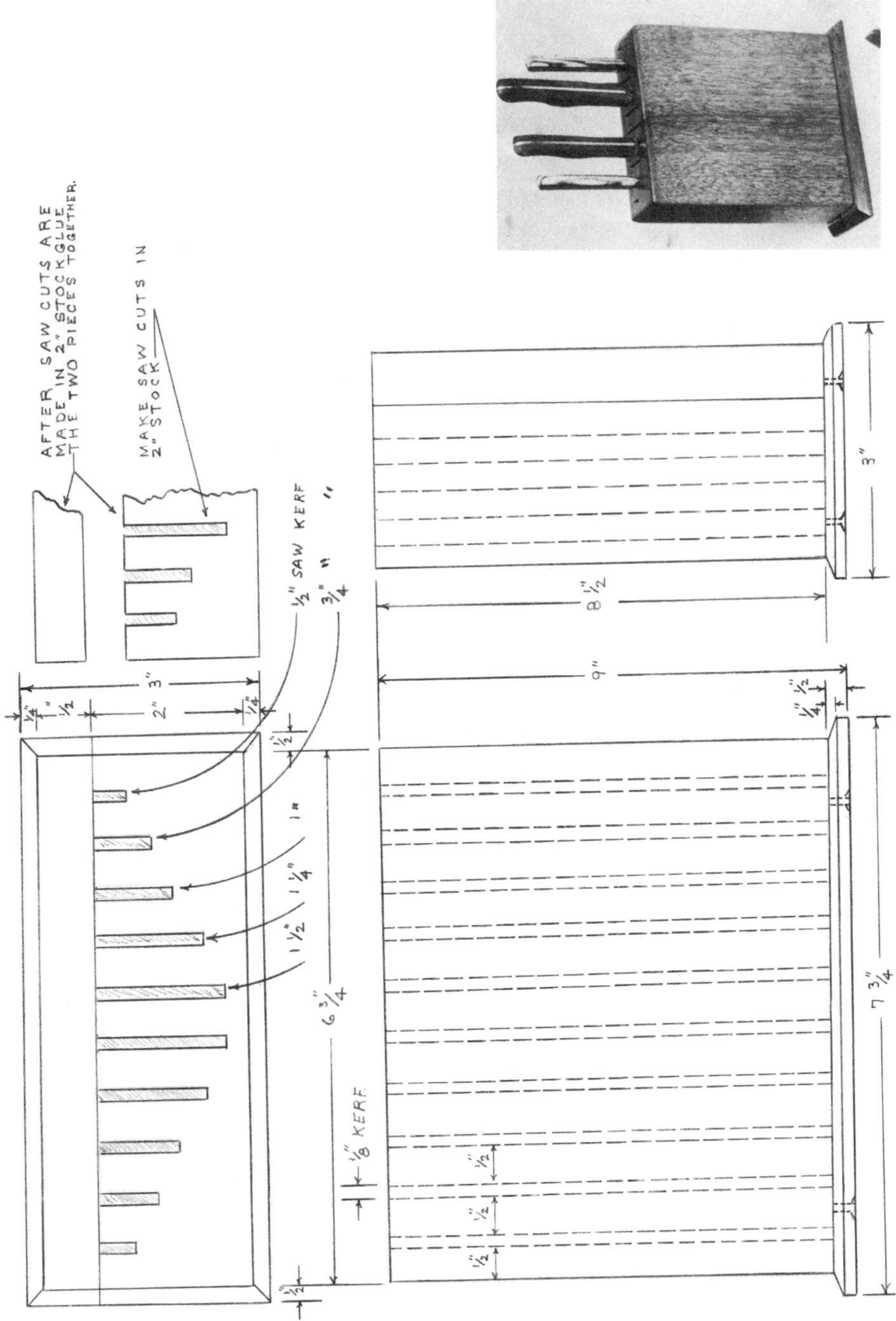

## Project 40

# COAT RACK

### Materials Required

*Wood of Your Choice*

2 **bases** 2½″ × 2½″ × 17″
4 **braces** ¾″ × 3″ × 6″
4 **hooks** ¾″ × 3″ × 6″
1 **center post** 2″ × 2″ × 16″
12 **wood screws** #10, 1½″ flathead
1 **wood screw** #14, 4″ flathead

### Procedures

1. Lay out and cut the stock to the size and design indicated on the drawing.
2. Lay out and cut a half-lap joint into the two base pieces. Fasten the base together by gluing and clamping the two pieces into the half-lap joint.
3. Drill a pilot hole through the center of the base and center post, then attach the center post to the base with a large screw.
4. Drill three ⅜-inch countersunk pilot holes into each brace, then fasten the four braces to the center post and base with #10, 1½-inch flathead wood screws. Cover all screw holes with ⅜-inch furniture buttons.
5. Drill two ⅜-inch countersunk pilot holes into each coat hook, then fasten the four coat hooks to the top of the center post with #10, 1½-inch flathead wood screws. Cover all screw holes with ⅜-inch furniture buttons.
6. Remove all traces of glue, then sand the project to a smooth surface.

# Coat Rack

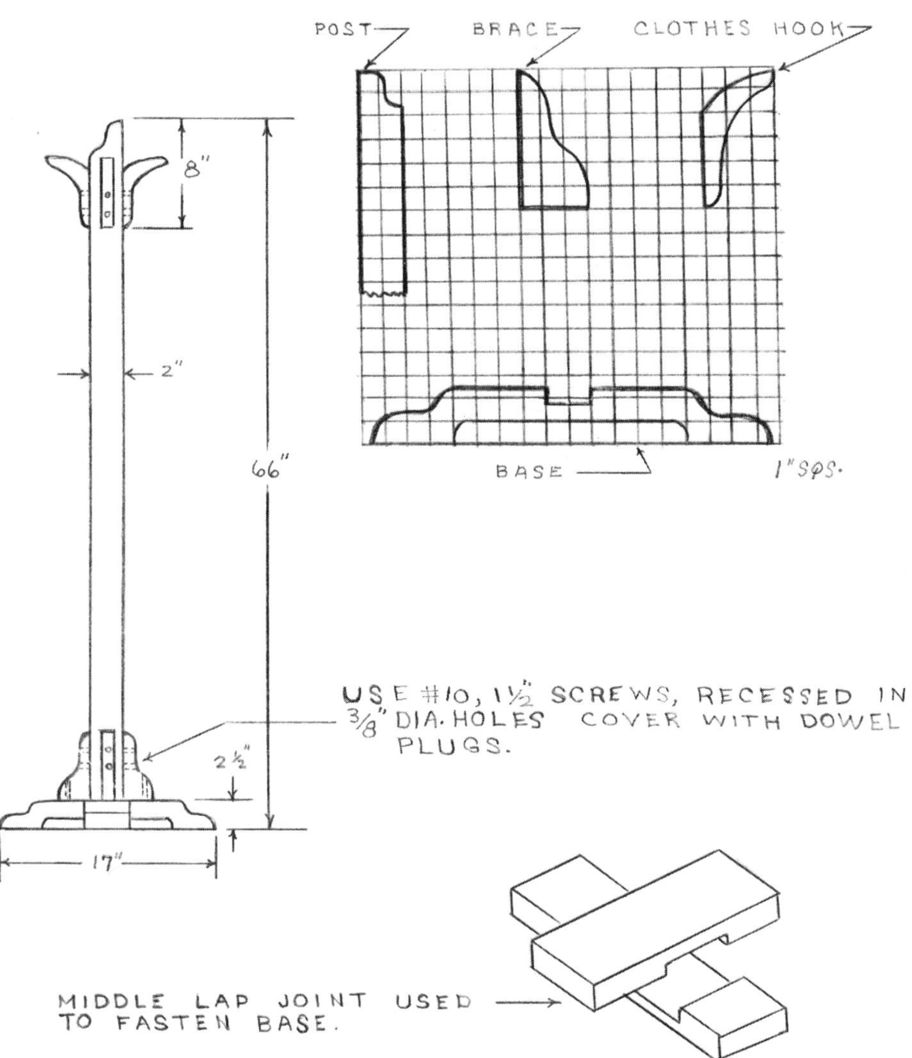


7. Select and apply the colored stain or paint of your choice. When dry, apply several coats of clear polyurethane finish, rubbing between each coat with fine steel wool or pumice and oil. To protect the finish apply a coat of paste wax and buff.

## Project 41

# DOUBLE BED

### Materials Required

*Wood of Your Choice*

**2 platform sheets** ¾″ × 26¾″ × 76″

*Bottom Frame:*
**2 sides** 2″ × 8″ × 68″
**2 ends** 2″ × 8″ × 43″
**2 cross braces** 2″ × 4″ × 43″
**2 bed sides** ¾″ × 8″ × 76″
**1 headboard** ¾″ × 16″ × 55″
**1 footboard** ¾″ × 12″ × 55″
**2 rims** ¾″ × 2″ × 74½″
**2 rims** ¾″ × 2″ × 53½″

### Procedures

1. Using a square, lay out all cuts to the size indicated in the drawing, then cut all of the parts to size.
2. Assemble the bottom frame of 2 inches by 8 inches, using glue and #12 finishing nails to hold the butt-jointed corners together. #12, 3-inch flathead screws can be used in place of nails. Check with a square to see that the corners are at a 90-degree angle. Nail the cross braces to the frame with #12 finishing nails.
3. Assemble the four pieces that make the frame to be placed around the mattress. The corners are to be simple butt joints fastened with glue and #10, 1½-inch flathead wood screws. There should be 3 to 4 screws per butt joint. The pilot holes for the screws should be counterbored so that they can be plugged with dowels or furniture buttons.

# Double Bed

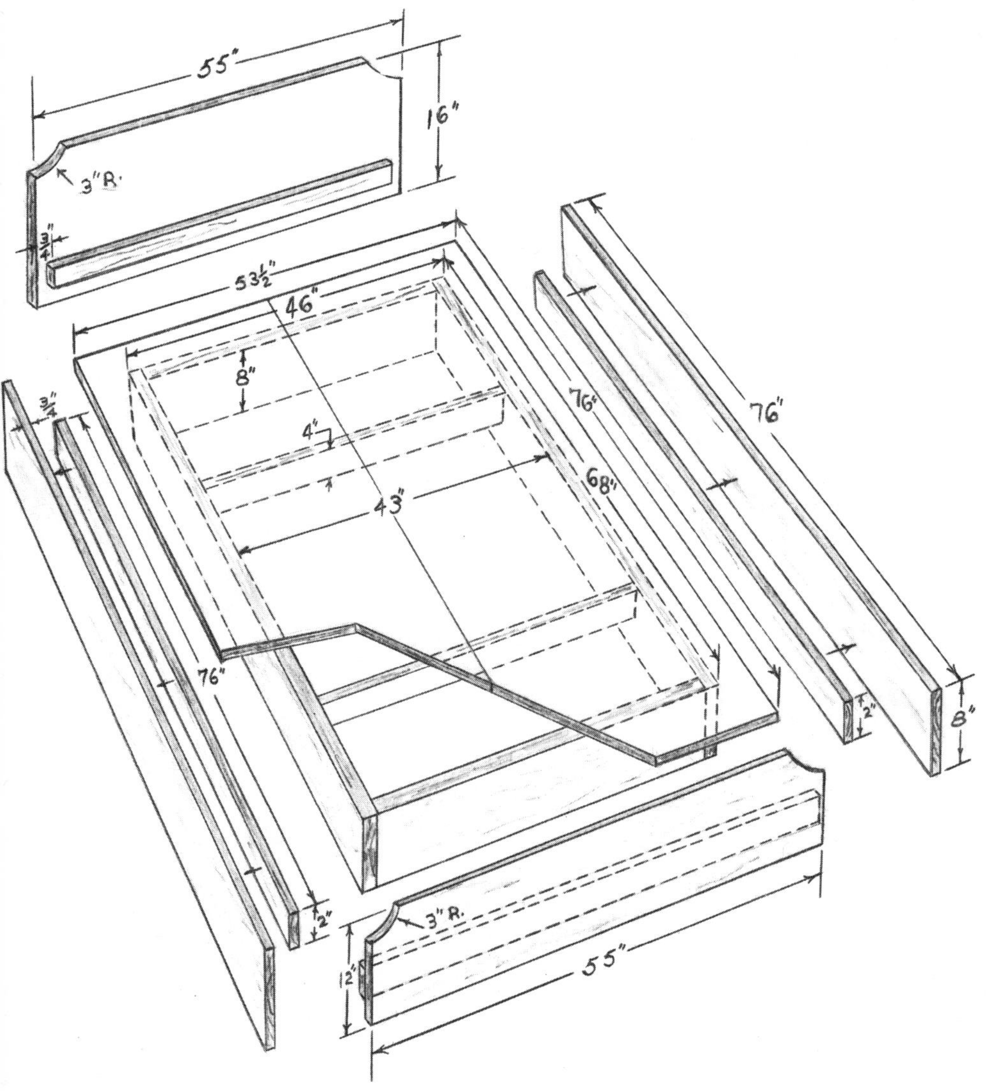

4. Fasten the ¾-inch by 2-inch pieces of stock to the lower inside edge of the outside frame to create a rim.

   Drill countersunk pilot holes approximately 18 inches apart, then attach the pieces with glue and #9, 1¼-inch flathead wood screws.

5. Set the two platform sheets into the frame, setting them on top of the ¾-inch by 2-inch rim.

   Drill countersunk pilot holes spaced approximately 24 inches apart, then fasten them to the ¾-inch by 2-inch rim with #10, 1¼-inch flathead screws.

6. Plug all screw holes, then sand the bed to a smooth finished surface. Clean the dust off, then apply the stain or paint of your choice. When completely dry, apply several coats of clear polyurethane finish, rubbing between coats with fine steel wool or #320 emery paper. Preserve the finish with a coat of paste wax.

## Project 42

# BOOK/ALBUM RACK

### Materials Required

*Wood of Your Choice*

**2 sides** ¾″ × 8″ × 12″
**2 rails** ½″ × 3″ × 20″

### Procedures

1. Lay out and cut all materials to the size and design indicated in the drawing.
2. Cut the ⅜-inch by 2¾-inch slots in the two sides to receive the rails.
3. Cut two ¾-inch blind dadoes to form the notches in each rail. The depth should be ⅛ inch by 2¾ inches long. The blind dadoes can be cut on the table saw with a dado cutter, then they should be trimmed with a chisel.
4. Fasten the rails to the sides by applying glue to the dado notches, then slide the notches into the cutout slots in each side piece.
5. Clean off all traces of glue and sand the rack to a smooth surface.
6. Apply several coats of clear lacquer or polyurethane. Rub between each coat with fine steel wool or pumice and oil. To preserve the finish apply a coat of paste wax and buff.

# Book/Album Rack

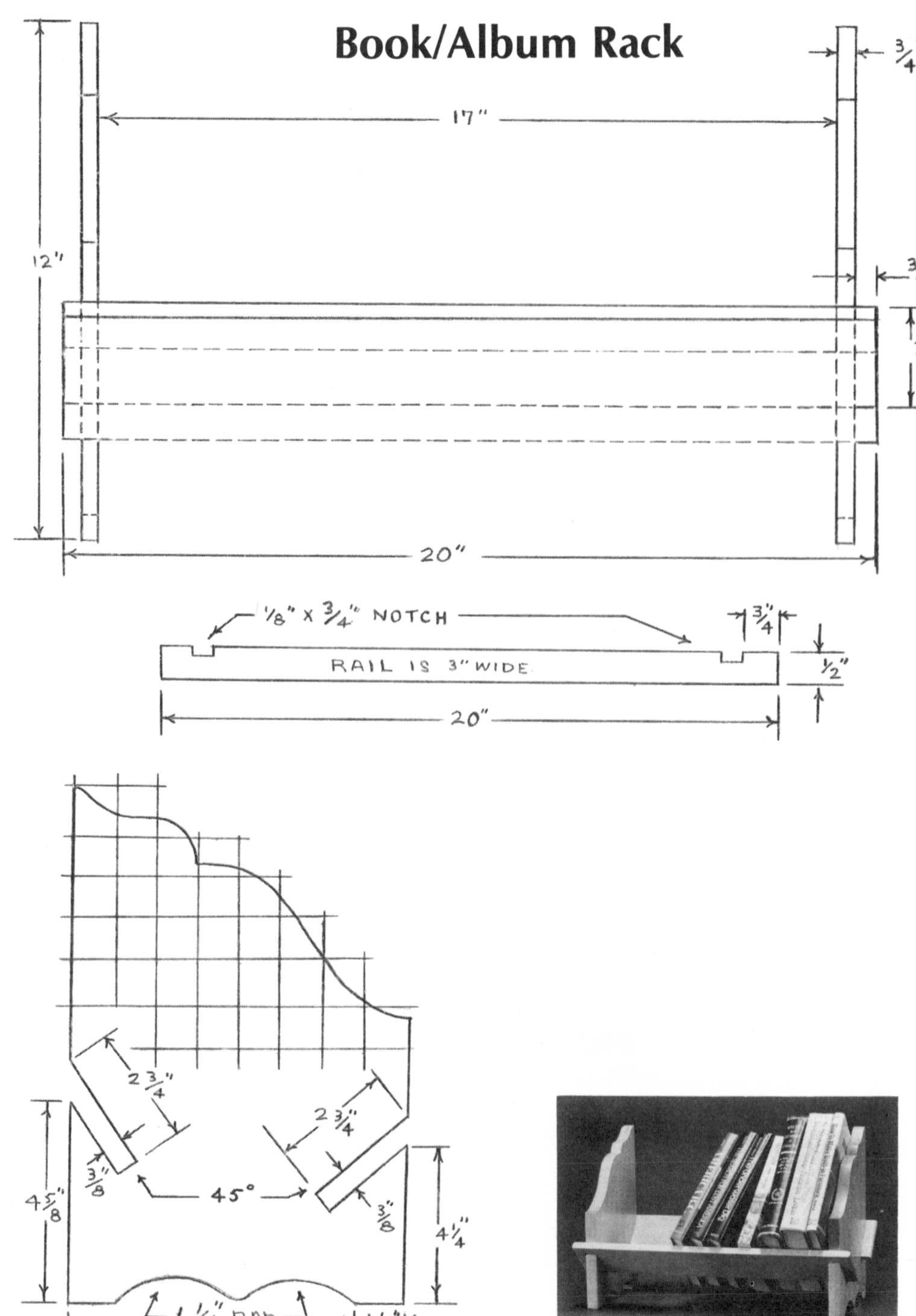

## Project 43

# BUNK BEDS

### Materials Required

*Wood of Your Choice*

**6 vertical bed posts** 2″ × 3″ × 72″

*Top Rail Parts:*

**1 piece** 2″ × 4″ × 18″
**1 piece** 2″ × 4″ × 48″
**2 pieces** 2″ × 4″ × 43″
**1 piece** 2″ × 4″ × 80″
**18 dowels** 1″ dia. × 10½″
**5 ladder rungs** 1″ dia. × 14″
**2 closet poles for hanging clothes** 1¼″ dia. × 46″

*Upper Bed Frame:*

**2 sides** 2″ × 6″ × 96″
**2 ends** 2″ × 6″ × 40″
**2 inner frames** 1″ × 2″ × 77″
**2 inner frames** 1″ × 2″ × 44″

*Lower Frame:*

**2 sides** 2″ × 6″ × 80″
**2 ends** 2″ × 6″ × 40″
**2 inner frames** 1″ × 2″ × 77″
**2 inner frames** 1″ × 2″ × 44″
**2 mattress panels** ¾″ × 40″ × 77″

*Note:* The materials given are geared to a mattress size of 6 inches by 39 inches by 76 inches. The bed frame allows a 1-inch space for blankets in both the width and the length. If the beds are to take a larger or smaller mattress, change the dimensions accordingly.

# Bunk Beds

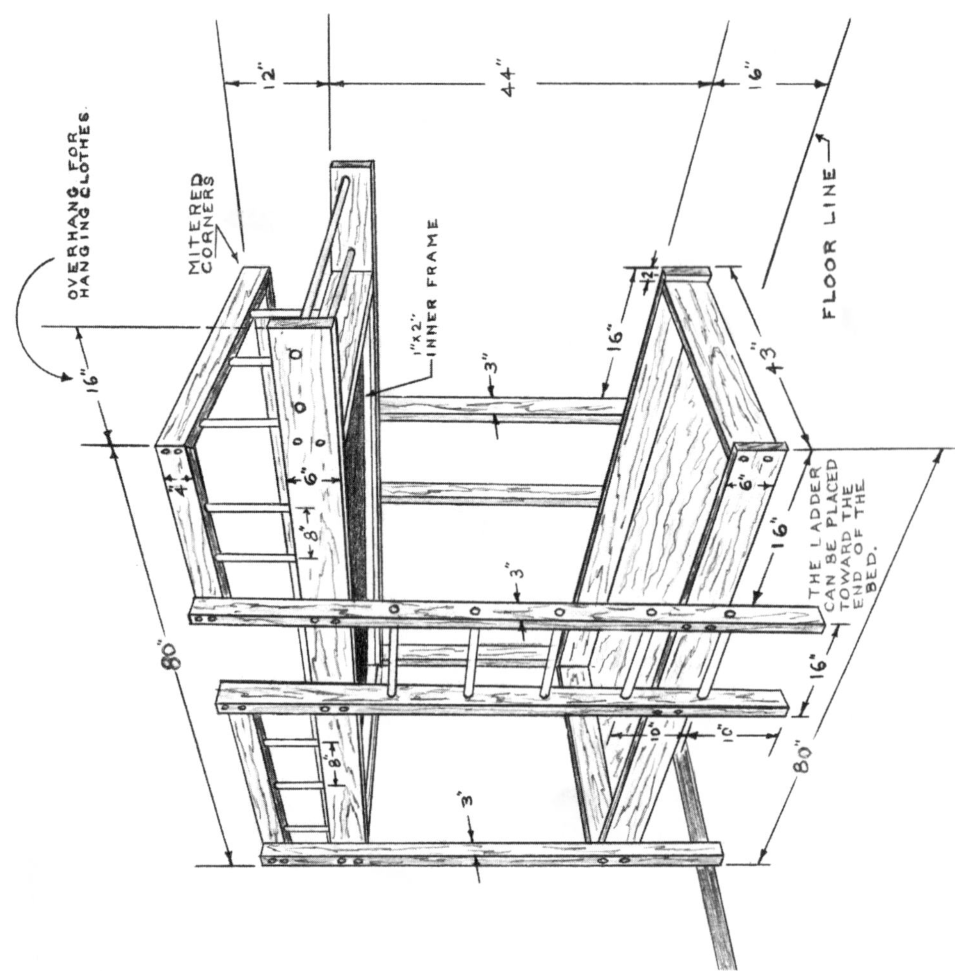
OVERHANG FOR HANGING CLOTHES
MITERED CORNERS
1"x2" INNER FRAME
FLOOR LINE
THE LADDER CAN BE PLACED TOWARD THE END OF THE BED.
12"
44"
16"
43"
80"
80"

## Procedures

1. Cut all of the pieces to the sizes indicated in the drawing and in the list of materials. Use a square to lay out and check cuts for a 90-degree angle.

2. To assemble the top bed frame use a spade bit to bore the matching 1½-inch-diameter holes to receive the closet poles for hanging clothes.

   Insert the poles into the holes.

   Locate and place the 2-inch by 6-inch end pieces between the 2-inch by 6-inch side pieces and drill counterbored pilot holes through the sides and into the ends of the frame. Be sure the counterbored pilot holes are large enough to handle #12, 3-inch flathead screws. Use at least two screws for each joint. Assemble the lower frame in the same manner.

3. To construct the ladder lay out on the 2-inch by 3-inch pieces all of the matching 1-inch-diameter holes for the rungs at a 10-inch spacing. Use a spade bit to bore the holes completely through.

   Apply glue to the dowel ends, then insert the dowels into the holes of one of the 2-inch by 3-inch pieces. Fasten the second 2-inch by 3-inch piece to the dowels by using a hammer and a wooden scrap block to prevent any dents to the wood. Be sure when boring the holes to keep them aligned so the rungs will be parallel.

4. To secure the bed frames to the vertical post and ladder, drill all of the counterbored holes for the #12, 3-inch screws through each vertical member.

Lay the two frames on their side, then fasten the ladder and end post to the frames, making sure they are positioned according to the dimensions given in the drawing.

With help from a friend turn the assembly to the opposite side and secure the three vertical posts in the same manner. Reinforce the bed frame to the post assembly by fastening screws from the inner face of the frame on through to the post.

5. Miter the four corners for the top rails, then fasten them together by drilling counterbored pilot holes and screwing with #12, 3-inch screws.

   Bore 1-inch-diameter holes that are spaced approximately 6 to 10 inches apart, then insert 1-inch dowels with glue.

6. Use 1-inch by 2-inch inside framing to hold the mattress panels in place.

   Secure the 1-inch by 2-inch inner frame to form a rim. The bottom edge of the 1-inch by 2-inch pieces should be flush to the bottom edge of the 2-inch by 6-inch pieces. Use glue and #6 nails to fasten.

   Set the ¾-inch-thick pieces on top of the 1-inch by 2-inch rim. Secure the panels to the rim with #10, 1½-inch screws spaced approximately every 18 inches.

7. Use hardwood dowel plugs to fill all counterbored holes, then sand the bunk beds smooth. Apply several coats of clear polyurethane finish, rubbing between coats.

Project 44

# FULL-LENGTH DRESSING MIRROR

## Materials Required

*Wood of Your Choice*

**2 mirror frames** (top and bottom) 1″ × 2½″ × 15″
**2 sides** 1″ × 2½″ × 52″
**1 backing for mirror** ¼″ × 19″ × 51″
**2 legs** 1½″ × 7″ × 16″
**2 posts** 1½″ × 2½″ × 36″
**1 cross-rail** 1½″ × 2½″ × 20½″
**2 wood screws** #12, 2½″ flathead
**2 furniture buttons** ½″ dia.
**1 mirror** ¼″ × 16″ × 48″ The mirror can be purchased from and installed by your local glass dealer.
**1 dowel** ⅜″ dia.
**1 dowel** ½″ dia.

## Procedures

1. Begin by cutting to width and length all of the pieces necessary to construct the mirror frame. Use a square to lay out and check for accuracy.
2. With a table saw or an electric router, cut rabbets on the four frame pieces to receive the mirror.
3. Lay out and drill matching ⅜-inch-diameter holes to receive the ⅜-inch-diameter by 2½-inch dowels for the corner joints. Make sure the holes are aligned. Use a dowel jig for accuracy. Glue the corners. Use barclamps to apply pressure to the glued corners. Check squareness with a framing square.

# Full-length Dressing Mirror

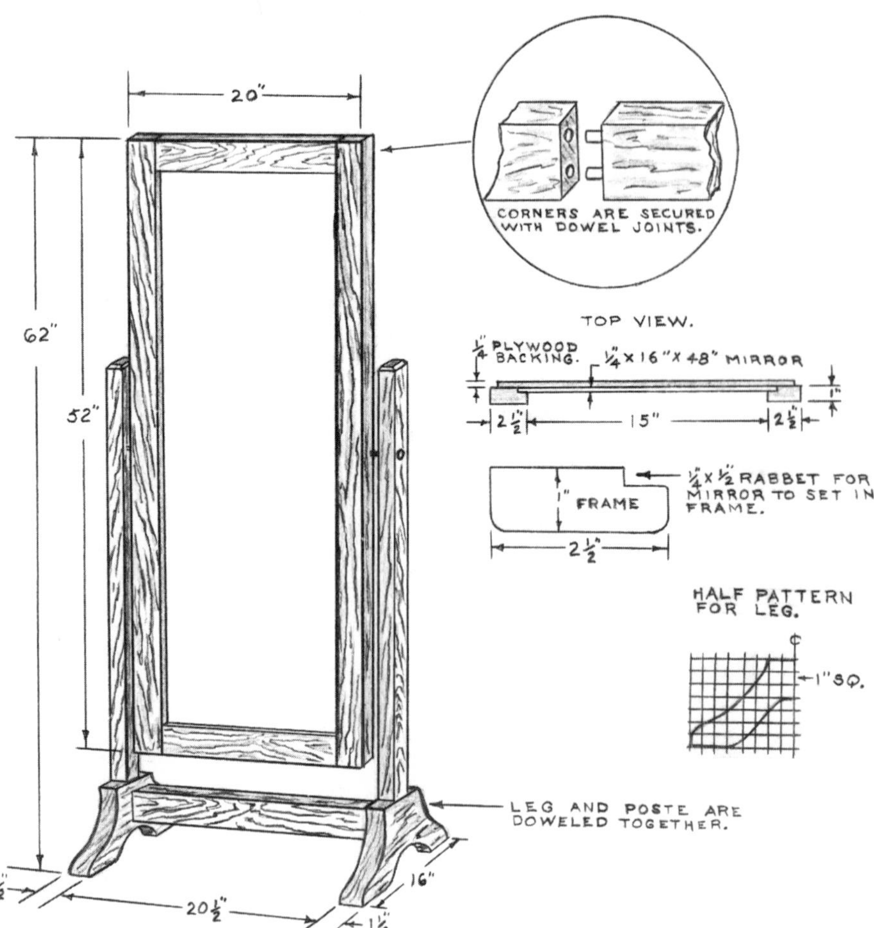


4. Lay out and cut the legs, post, and cross-rail to the size and shape indicated in the drawing.
5. For each leg-and-post assembly locate and drill two matching ½-inch-diameter by 2-inch holes for a doweled joint. Secure the post to the base leg by inserting the ½-inch dowels

with glue. Apply even pressure with bar clamps.

6. Fasten the cross-rail to the base legs by drilling two counterbored pilot screw holes through each leg and into the rail then fastening with #10, 3-inch flathead screws. Cover counter-bored holes with hardwood plugs.

7. To secure the mirror frame to the post start by drilling a 3⁄16-inch-diameter hole completely through the post. The hole should be located exactly 8 inches down from the top edge of the post. Next drill a ½-inch counterbored hole over the 3⁄16-inch-diameter hole. The counter-bored hole should be ¾-inch deep.

   Locate the 26-inch center mark on the edge of the mirror frame, then drill a matching ⅛-inch anchor hole.

   Fasten the frame to the post by inserting #12, 2-inch flathead screws. The screws will hold secure in the anchor hole in the frame and swivel freely in the counterbored pilot hole in the post. Cover the counterbored holes with hardwood plugs.

   *Note:* To keep the mirror frame from rubbing the post insert a small nylon washer between the two pieces.

8. Fill holes, scrape off all traces of glue, then sand the wood to a final smooth surface. Select and apply the stain or paint of your choice. When dry, apply three coats of poly-urethane clear finish. Rub between coats with fine steel wool or #320 emery paper. Preserve the finish with an occasional coat of lemon oil.

9. Have your glass dealer cut and set the mirror into the frame, then apply the ¼-inch backing with 1-inch finishing brads.

## Project 45

# WALL SHELF

### Materials Required

*Wood of Your Choice*

**2 sides** ¾″ × 6½″ × 25½″
**1 top shelf** ¾″ × 4⅜″ × 24¼″
**1 middle shelf** ¾″ × 5½″ × 24¼″
**1 bottom shelf** ¾″ × 6½″ × 24½″
**1 leather strip for hanging shelf** ⅛″ × ¼″ × 30″

### Procedures

1. Lay out and cut all pieces to the suggested size and design.
2. Lay out and cut the dado and rabbet joints to a depth of ⅜ inch.
3. Assemble the shelves to the sides by gluing and clamping.
4. Plane the protruding edges of the shelves even to the sides of the wall shelf.
5. Drill two ¼-inch holes through the sides to hang the shelf.
6. Scrape off all traces of glue, then sand the wall shelf to a smooth surface.
7. Select and apply the stain or paint of your choice. When dry, apply several coats of clear finishing oil, rubbing between coats with fine steel wool or pumice and oil. Protect the finish by applying a coat of paste wax and buffing.

# Wall Shelf

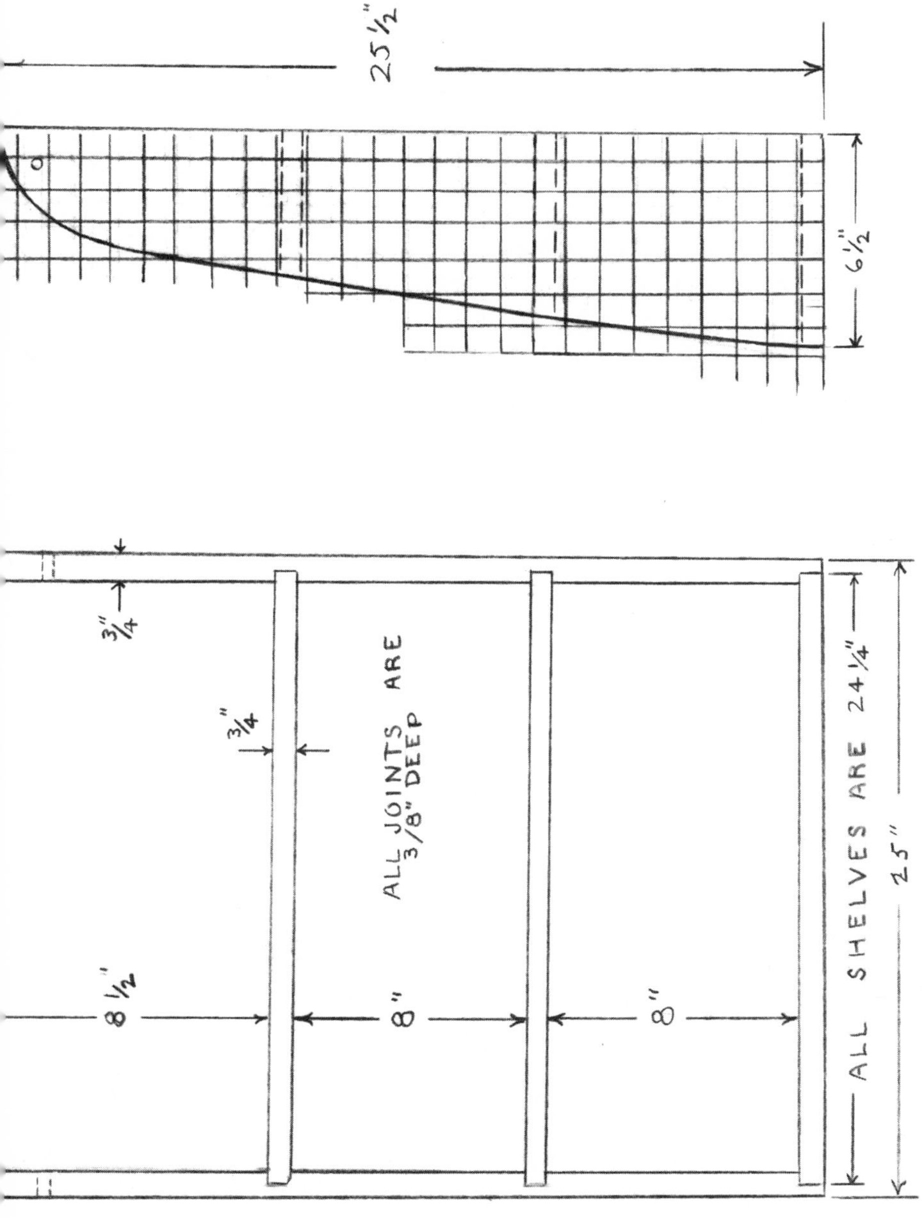

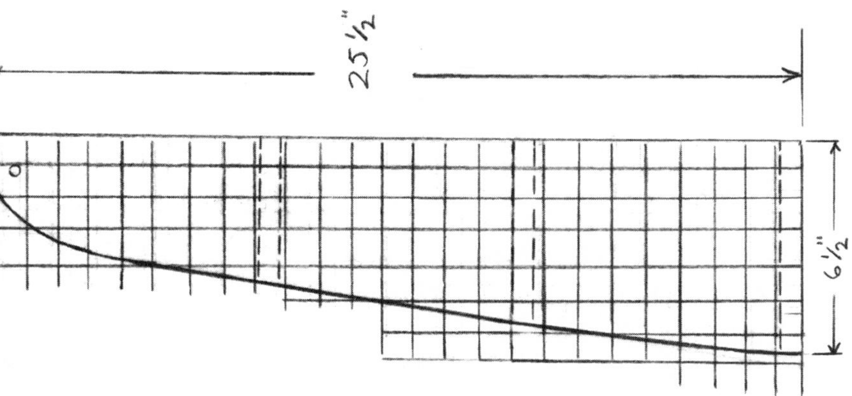

25 1/2"
6 1/2"
3/4"
3/4"
ALL JOINTS ARE 3/8" DEEP
ALL SHELVES ARE 24 1/4"
25"
8 1/2"
8"
8"

## Project 46
# DESK

### Materials Required
*Wood of Your Choice*

**1 top** 1″ × 22½″ × 48″
**2 sides for drawer cabinet** 1″ × 18½″ × 29″
**1 back** 1″ × 16″ × 29″
**8 drawer guides** ¾″ × 1½″ × 16½″
**3 drawer rails** ¾″ × 1¾″ × 16″
**1 front skirt** ¾″ × 4¾″ × 18″
**1 writing board** ¾″ × 15⅞″ × 16″
**1 writing board front** ¾″ × 2″ × 16½″

*Materials for One Drawer:*
**1 front** ¾″ × 16¼″ × 16½″
**1 back** ⅜″ × 5⅞″ × 15⅞″
**2 sides** ⅜″ × 5⅞″ × 17″
**1 bottom** ¼″ × 15⅛″ × 16¼″

*Materials for Leg Assembly:*
**2 pieces** 2″ × 2″ × 29″
**2 pieces** 2″ × 2″ × 14½″
**2 pieces** 2″ × 2″ × 19″
**1 foot rest brace** 2″ × 3″ × 24″

**hardware of your choice**

### Procedures

1. Edge-glue enough stock to meet the required widths for the top and cabinet.
2. Lay out and cut the two sides and the back for the drawer cabinet. Glue and nail or screw the assembly together. If screws are used, the pilot holes should be counterbored so that dowel plugs can cover the screws.

# Desk

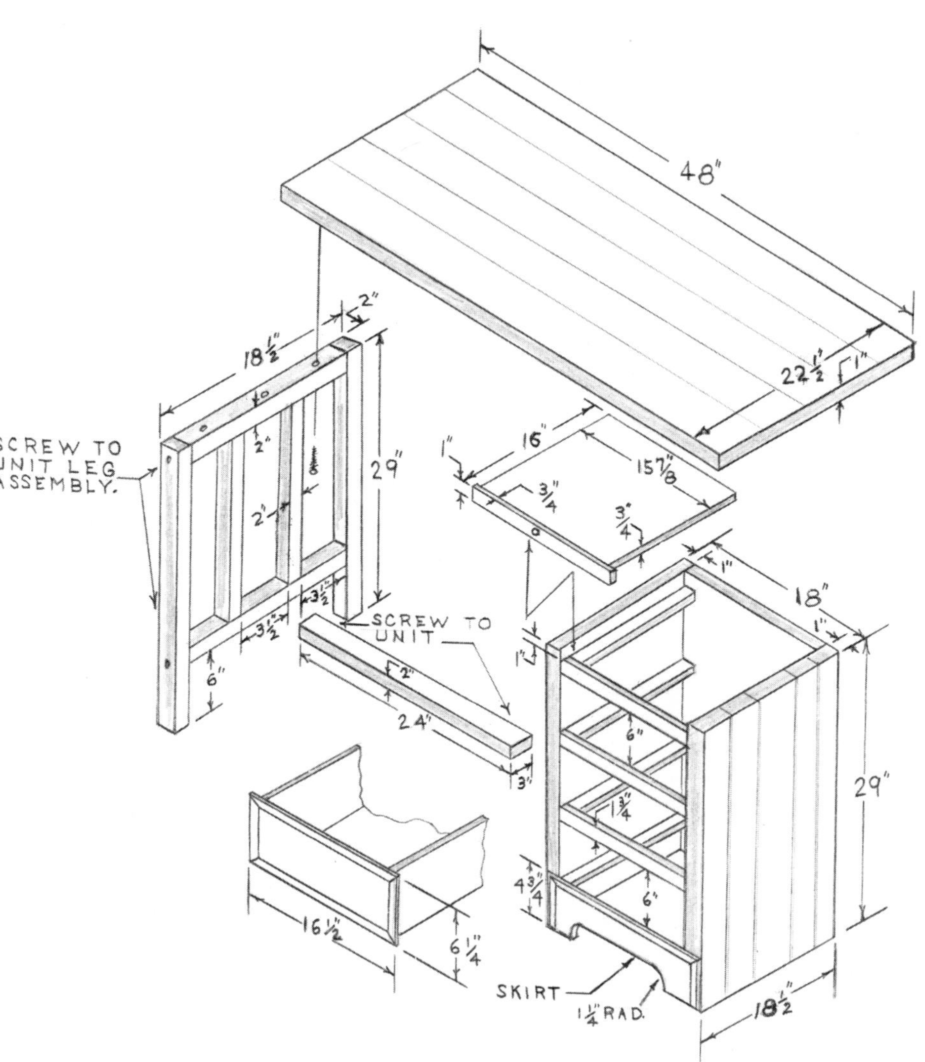
48"
22 1/2"
1"
18 1/2"
2"
29"
SCREW TO
UNIT LEG
ASSEMBLY.
2"
2"
3 1/2"
3 1/2"
6"
16"
1"
15 7/8
3/4"
3/4"
SCREW TO
UNIT
2"
24"
3"
1"
18"
1"
1"
6"
1 3/4
4 3/4"
6"
29"
16 1/2
6 1/4"
SKIRT
1 1/4" RAD.
18 1/2"

3. Lay out and cut the three front rails and the base skirt.

   Lay out and cut a ⅜-inch by ¾-inch chamfer on the three exposed edges of the skirt. Use a hand plane to complete the chamfer.

   Fasten the four pieces to the front of the open assembly with glue and #6 finishing nails.

   The assembly is completed by cutting the drawer guides and fastening them to the sides with #8, 2¼-inch flathead wood screws. At this point the assembly has three equal spaces to receive the three large drawers.

4. Lay out and cut the stock to the suggested size for the foot rest and the leg assembly.

   Attach the leg assembly by locating and drilling ⅜-inch-diameter counterbored pilot holes and then fastening the five pieces together with glue and #10, 3-inch flathead wood screws.

5. Attach the top to the drawer assembly and leg assembly with #10, 3-inch flathead wood screws spaced evenly. Attach the foot rest to the leg assembly and drawer assembly with #10, 3-inch flathead wood screws. Be sure to counterbore the pilot holes so they can be plugged.

6. Construct three drawers with a ¼-inch by ¾-inch chamfer on the front of each drawer. The drawer fronts should have a ¼-inch by ¾-inch rabbet cut to receive the sides of the drawer and then a ¼-inch by ⅜-inch rabbet cut on the top and bottom so that there will be a ⅜-inch overlap all around. (Refer to the section on drawer construction in the front of the book.)

7. Set and fill all nail holes. Cover the screw

heads with hardwood plugs, then sand the desk to a smooth surface.

Stain to your choice of color, then apply several coats of lacquer or varnish. Rub the project down with fine steel wool or pumice and lemon oil. Apply a final coat of paste wax to protect the finish.

## Project 47

# TILT-TOP DESK

### Materials Required

*Wood of Your Choice*

**2 sides** ¾″ × 32″ × 28″
**1 top** ¾″ × 29″ × 56½″
**1 back** ⅜″ × 12″ × 56½″
**1 drawer case** (overall size) 12″ × 15″ × 28″ All stock is ¾″ thick except for divider, which is ½″.
**flush front drawer** (overall size) 4 15/16″ × 13⅜″ × 26½″
**3 dowels** 1¼″ dia. × 5′
**2 lid supports or 12″ casement windows**
**2 locking guides**
**2 leather strips or pipe clamps** ⅛″ × 2″ × 3″
**4 drawer pulls of your choice**

### Procedures

1. Determine if the desk is to be made from solid core birch plywood or standard hardwood stock. If the standard stock is to be used, rough cut the stock then use bar clamps to edge-glue up enough pieces to meet the required width for all parts.
2. Cut all pieces of the desk to the size shown in the drawing. Use a carpenter's square to mark and check for accuracy, then sand all of the pieces in preparation for assembly.
3. Lay out and bore matching 1¼-inch-diameter holes with a spade bit. Bore completely through the sides to receive the dowels which the top is set on. Use scrap wood on the back of the stock to prevent the wood from splitting out when boring with the large spade bit.

# Tilt-top Desk

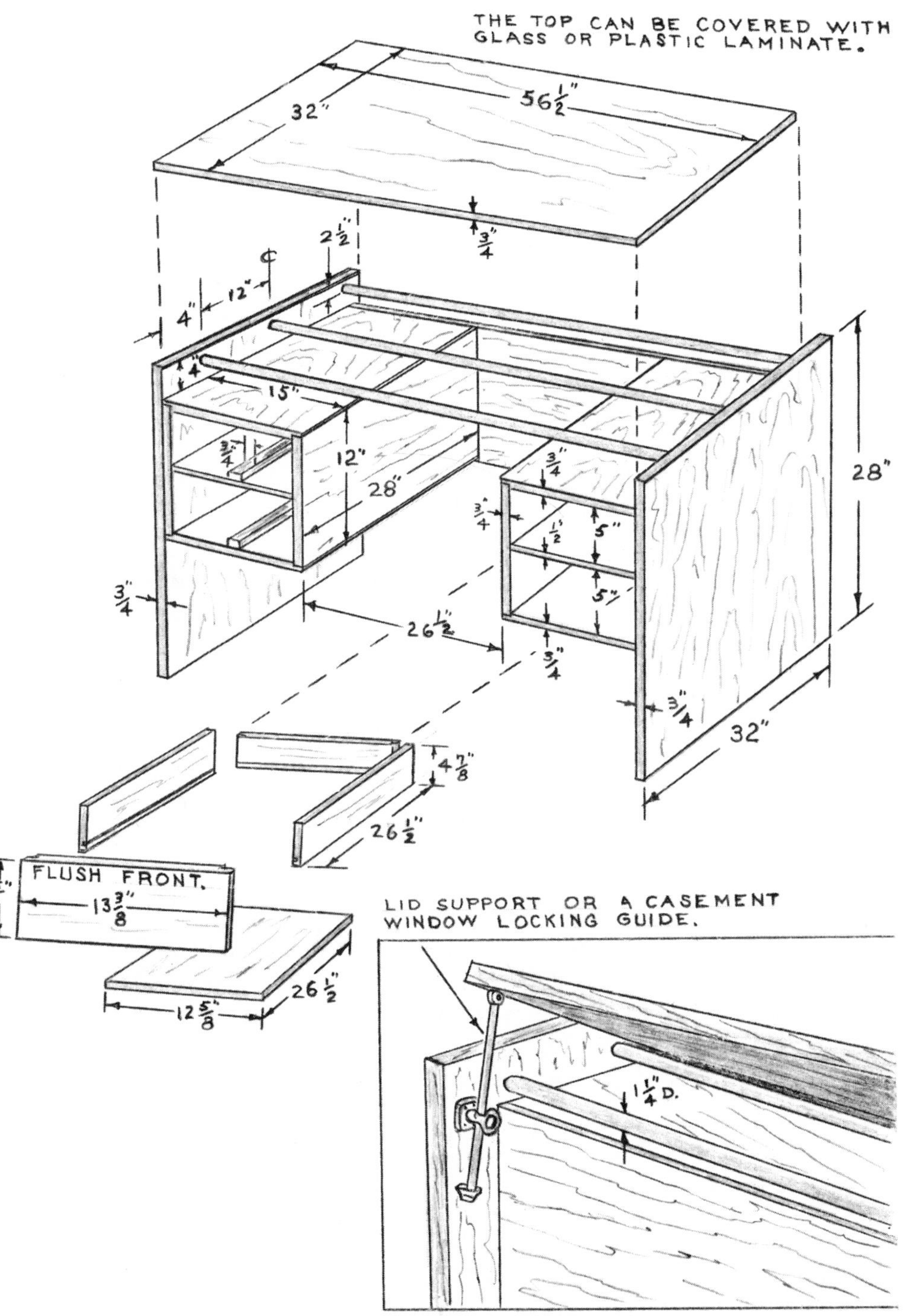

4. Construct the two drawer cases to the dimensions given in the drawing. The corner joints can be rabbets or 45-degree miter joints. The drawer divider can be nailed flush or it can be recessed into ⅜-inch by ¾-inch dadoes with glue. Leave the back of the cases open, as they will be covered when the back of the desk is installed.
5. Cut the three dowels to 58 inches, then spread a little glue into the dowel holes. Fasten the three dowels in the prebored holes, one side first. Then use a wooden mallet to pound the other side onto the dowels. Protect the wood by holding a scrap wood block against the legs as you pound the dowels in. Use a wet cloth to wipe away excess glue.
6. Secure the two drawer cases to the inner face of the two sides by gluing and clamping with wooden screw clamps. Be sure that the cases are positioned squarely and that they are completely aligned to each other.

   *Note:* The desk in the drawing shows the drawer cases flush to the front edge. It also is attractive if the cases are recessed 1 inch to 2 inches. Use #4 finishing nails to fasten the ⅜-inch back to the rear of the two drawer cases.
7. Construct the four flush front drawers to the dimensions given in the drawings. Refer to the section in the front of the book on drawer construction and center guide hints.
8. Set the top in place on top of the three dowels.

   Secure the top to the dowel located at the front of the desk by fastening two 2-inch-wide by 3-inch-long leather straps around the underside of the dowel and to the underside of the top by tacking them in place. An option is to use pipe clamps around the dowel, screwing into the underface of the top. This type of fastening device allows the top to pivot open and closed.

Secure the back of the top to the rear of the desk sides by using window glides or lid supports, as shown in the drawing.

9. Fill all holes, then give the desk a final sanding to a smooth surface. Select and apply the stain of your choice, then give three coats with #400 emery paper or fine steel wool. Preserve the finish with an occasional coat of lemon oil.
10. Locate and install the drawer pulls of your choice.

## Project 48

# OVAL MIRRORS

## Materials Required

*Wood of Your Choice*

*Oval Frame:*
**1 frame** ¾″ × 15″ × 20″
**1 oval mirror** ⅛″ × 10½″ × 16″

*Rectangular Frame:*
**1 frame** ¾″ × 15″ × 21″
**1 oval mirror** ⅛ ″ × 9″ × 16″

*Note:* The oval mirror can be cut to size at your local glass supply store. Oval mirrors can be purchased precut from your local craft supply store.

## Procedures

1. Glue and clamp the stock to the required width. Select and lay out the design of the mirror on the glued-up board.

2. Cut the board to the selected design. To cut out the oval design bore a ⅜-inch-diameter hole, then insert the blade of a sabre saw to do the cutting.

3. With a router and rabbet bit cut a 3⁄16-inch by ⅜-inch rabbet recess on the back side of the oval cutout to receive the oval mirror.

4. Scrape off all traces of glue. File and sand the frame to a smooth surface.

# Oval Mirrors

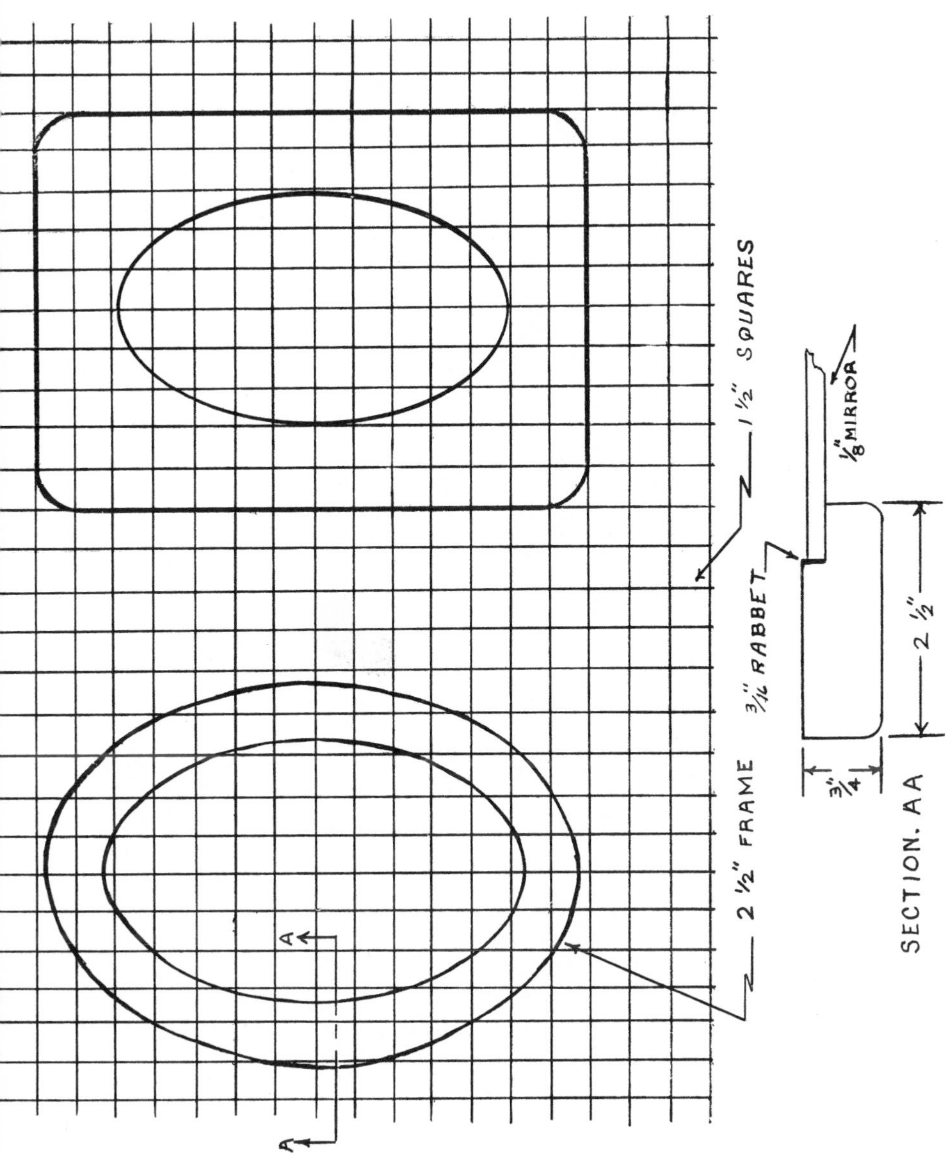

5. Select and apply the stain or paint of your choice. When dry, apply several coats of clear polyurethane finish, rubbing between coats with fine steel wool or pumice and oil. To protect the finish apply a coat of paste wax and buff.
6. Fasten the mirror into the rabbet recess with glass points placed approximately 5 inches apart.

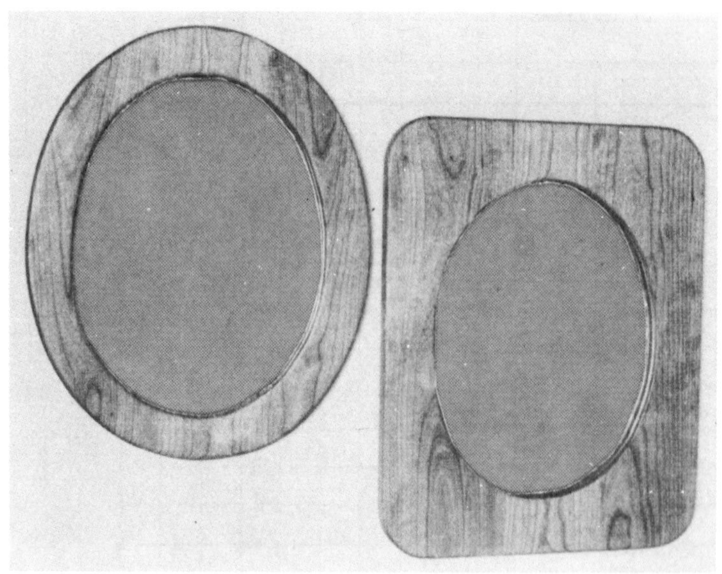

## Project 49

# LIVING ROOM COUCH

### Materials Required

*Wood of Your Choice*

**1 back** 1″ × 27″ × 70″
**2 sides** 1″ × 28″ × 37″
**1 seat** ¾″ × 17½″ × 73″
**1 front rail** 1″ × 4″ × 73″
**2 cleats** 1″ × 2¾″ × 15½″
**1 front inner rail to seat** 1″ × 2¾″ × 73″
**1 back rail to seat** 1″ × 3½″ × 70″

### Procedures

1. Start laying out and cutting the back of the couch to size. The two end edges of the back should be beveled to approximately 8 to 10 degrees so that when the sides of the couch are attached to the back, they will fan out approximately 3 inches. The beveled cut is made by tilting the blade of a table saw to 8 degrees, or by hand-planing to the correct angle.
2. Lay out and cut the remaining pieces to the size and design indicated in the drawing. Be sure to cut the ends of the seat panel at a taper to match the two sides that will fan out approximately 3 inches.
3. File and sand the armrests, then glue and clamp them to the sides of the couch.
4. Drill four countersunk pilot holes through each tapered cleat, then fasten them to the inner face of each side piece. The cleats should be attached approximately 2 inches back from the front edge of the side pieces.

# Living Room Couch

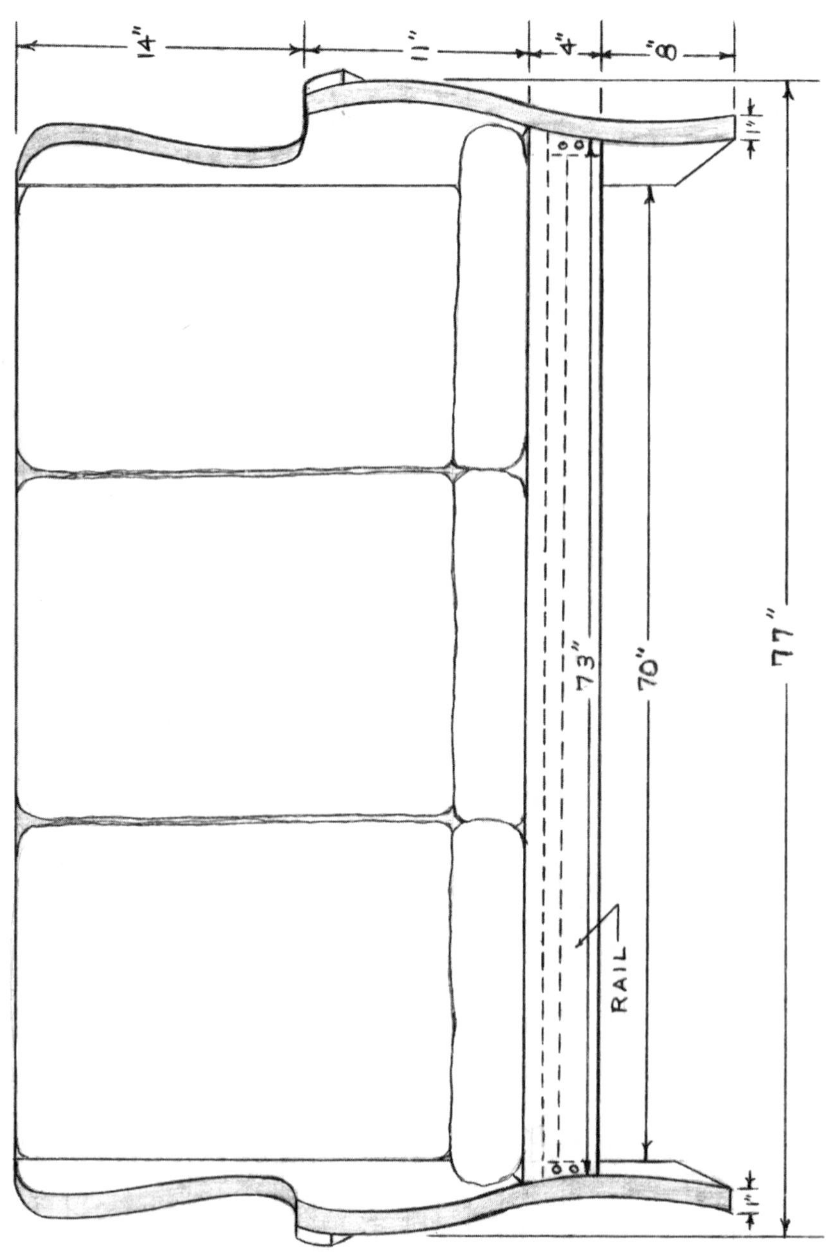

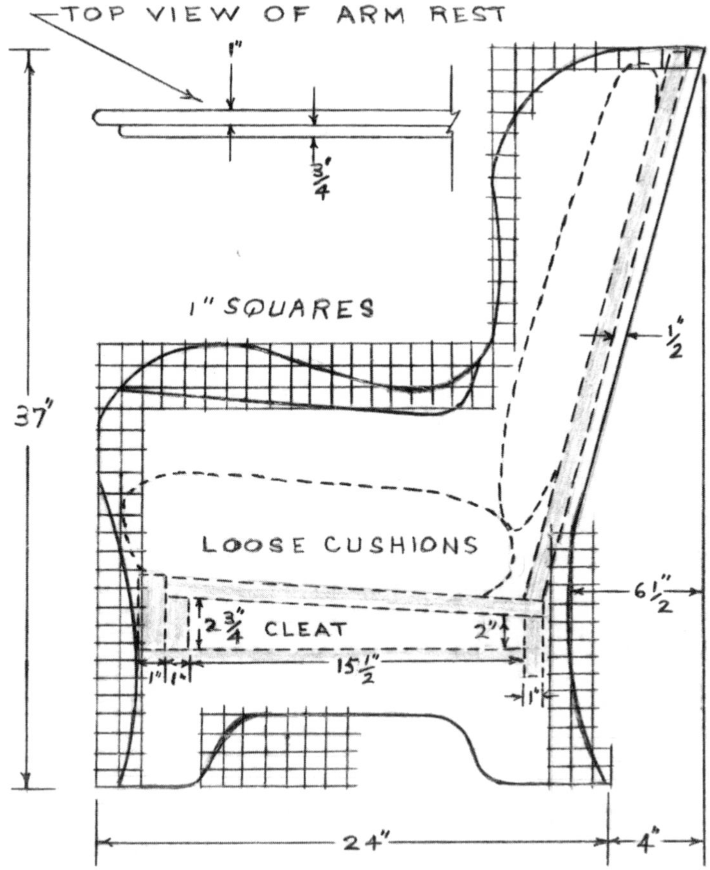


**5.** To finish the inner frame for the seat, glue and clamp a 2¾-inch by 73-inch piece of stock to the back of the front rail. The bottom edge of this piece should be placed even with the bottom edge of the front rail to form a rim so the seat can be set into place.

Screw these two pieces into the front edge of the side cleats.

Use two counterbored pilot holes through each end of the front rail and fasten with #12, 3-inch wood screws.

Fasten the 3½-inch by 70-inch back rail to the back edge of the cleats with screws.

6. Place the seat panel onto the rails and cleats, then drill countersunk pilot holes approximately every 18 inches. Fasten to the inner frame with #10, 1½-inch flathead wood screws.
7. Attach the back of the couch above the seat and to the sides by locating and drilling six counterbored pilot holes through each side piece. Use #10, 1½-inch flathead wood screws. Cover with furniture buttons or dowel plugs.
8. File and sand the scrolled edges thoroughly, then plug all screw holes with furniture buttons or dowel plugs. Sand the entire couch to a smooth surface.
9. Select and apply the stain or paint of your choice. When dry, apply several coats of clear polyurethane, rubbing between coats with fine steel wool or pumice and oil. Preserve the finish with an occasional coat of paste wax.

## Project 50

# LIVING ROOM CHAIR

### Materials Required

*Wood of Your Choice*

**1 back** 1″ × 27″ × 19″
**2 sides** 1″ × 28″ × 36″
**1 seat** ¾″ × 17″ × 22″
**1 front rail** 1″ × 4″ × 22″
**2 cleats** 1″ × 2¾″ × 15½″
**1 inner rail to seat** ¾″ × 2¾″ × 20″
**1 back rail to seat** ¾″ × 1½″ × 19″
**2 foam rubber cushions**

### Procedures

1. Begin by cutting the back to size and shape. The two end edges of the back piece should be cut to a bevel of approximately 8 degrees, so when the sides of the chair are fastened to the back, they will fan out. The angled cut is made by tilting the blade of a table saw to 8 degrees, or by hand-planing to the 8-degree angle.
2. Lay out and cut the remaining pieces to the size and design indicated in the drawing.

   Glue and clamp the two ¾-inch armrests to the chair sides.

   Drill four pilot holes through each tapered cleat, then fasten each cleat with #10, 1½-inch wood screws. The cleats should be placed 2 inches back from the front edge of the sides.

   *Note:* Be sure to cut the ends of the seat panel at a taper to match the two sides that will fan out approximately 3 inches.

# Living Room Chair

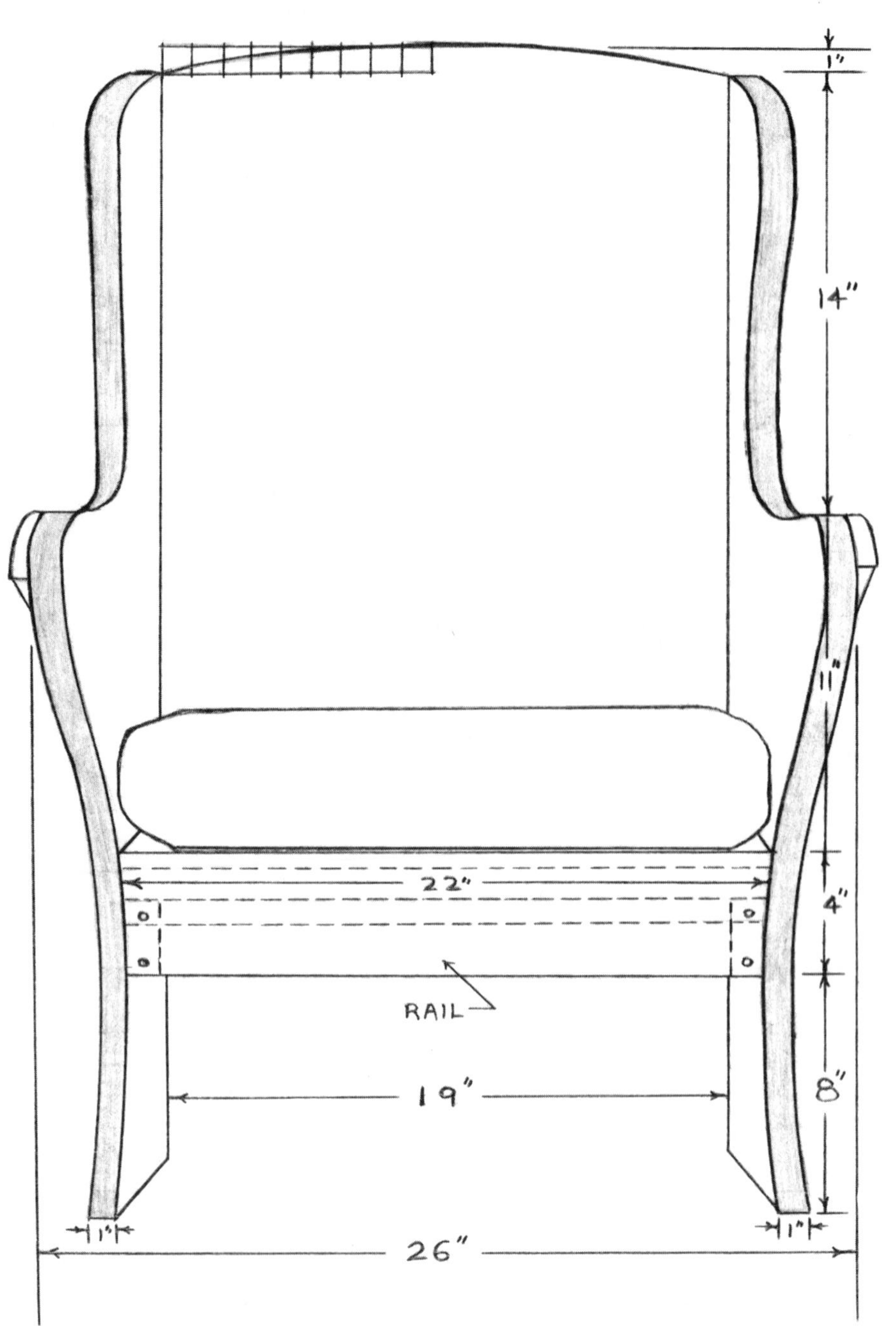

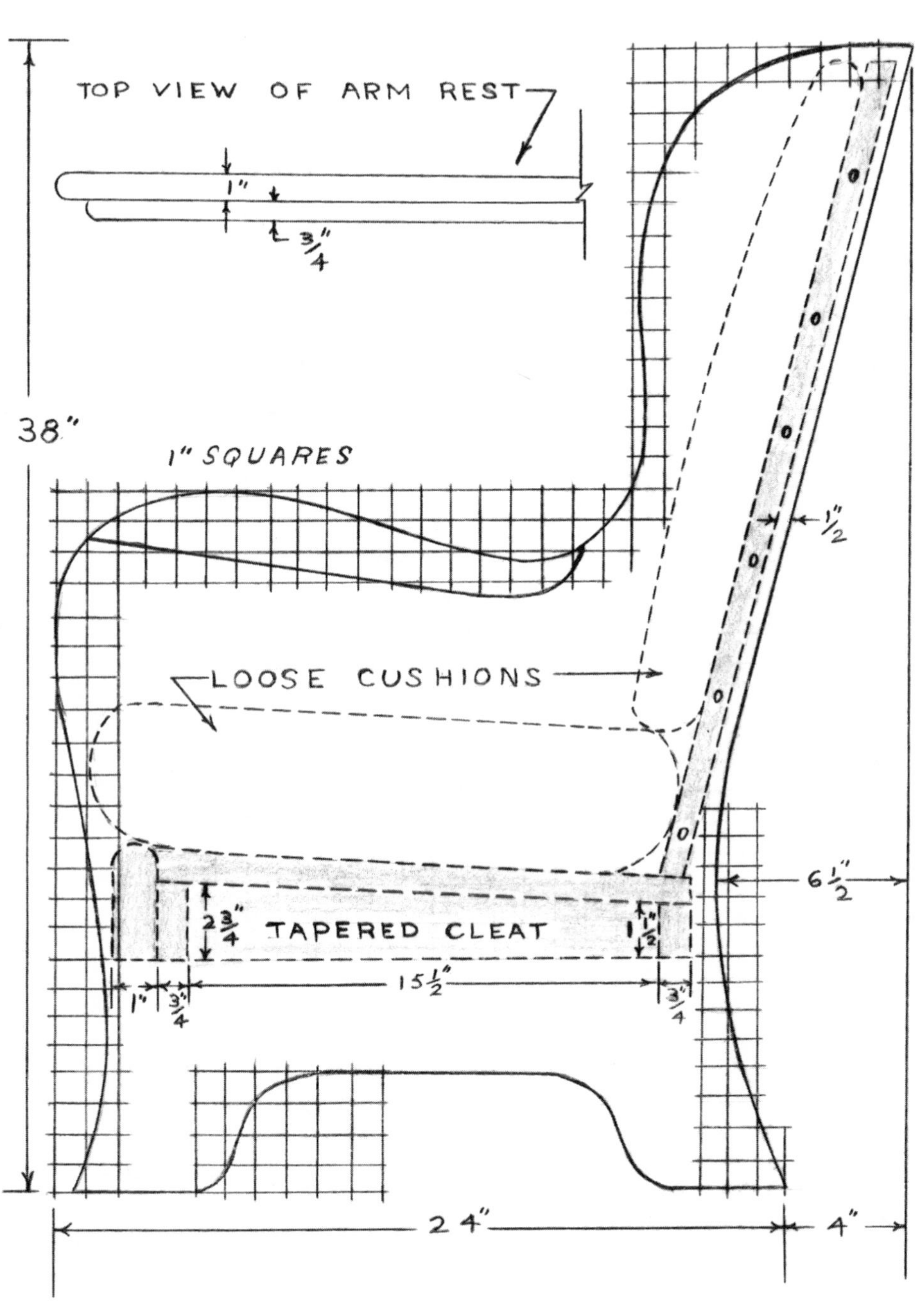
TOP VIEW OF ARM REST
1"
3/4"
38"
1" SQUARES
1/2"
LOOSE CUSHIONS
6 1/2"
2 3/4"
TAPERED CLEAT
1 1/2"
1"
3/4"
15 1/2"
3/4"
24"
4"

3. Glue and clamp a 2¾-inch by 22-inch piece of stock to the back of the 4-inch by 22-inch front rail. This piece is to form a rim so that the seat panel can be set in place.

   Attach the front rail to the front edge of the two side cleats by drilling two counterbored pilot holes through each end then installing #12, 2½-inch flathead wood screws.

   Fasten the back 1½-inch by 19-inch rail to the back edge of the cleats with screws.

4. Position the precut seat on the rails, then drill two counterbored pilot holes evenly spaced on each of the four sides. Fasten with #10, 1½-inch flathead wood screws.

5. Fasten the back of the chair above the seat and to the sides by locating and drilling six counterbored pilot holes through each side piece, then install #10, 1½-inch flathead wood screws.

6. File and sand the scrolled edges thoroughly, then plug all screw holes and continue to sand the entire chair to a smooth surface.

7. Select and apply the stain or paint of your choice. When dry, apply several coats of clear lacquer, rubbing between applications with fine steel wool or pumice and oil. Preserve the finish with an occasional coat of paste wax and buffing.

## Project 51

# TELEVISION CONSOLE

### Materials Required

*Wood of Your Choice*

**1 top** ¾″ × 18½″ × 31″
**2 sides** ¾″ × 18″ × 27¼″
**1 bottom shelf** ¾″ × 18″ × 29¼″
**2 front and back rails** ¾″ × 1¾″ × 30″
**2 cleats** ¾″ × 1″ × 16½″
**1 front skirt** ¾″ × 8″ × 31½″
**2 skirt sides** ¾″ × 8″ × 18¾″
**1 cove molding** ¾″ × ¾″ × 72″
**1 raised-paneled door** (overall size) ¾″ × 15⅜″ × 20¾″
**4 offset hinges** ⅜″
**2 magnetic catches**
**2 doorpulls of your choice**

### Procedures

1. Use bar clamps to edge-glue up enough stock to meet the required width for all pieces necessary.
2. Lay out and cut square (90 degrees) the top, bottom shelf, sides, and two rails.

   Make a rough cutout at the bottom of both sides so that the skirt design will fit over.

   Lay out and cut a ⅜-inch by ¾-inch dado into each side piece to receive the bottom shelf.
3. Secure the bottom shelf into the dadoes in the side pieces, then fasten the front and back top rails to the side pieces using glue and #6 finishing nails. For added strength glue a cleat flush to the top edge of the side pieces. The cleats should be ¾ by 1 by 16½ inches.

# Television Console

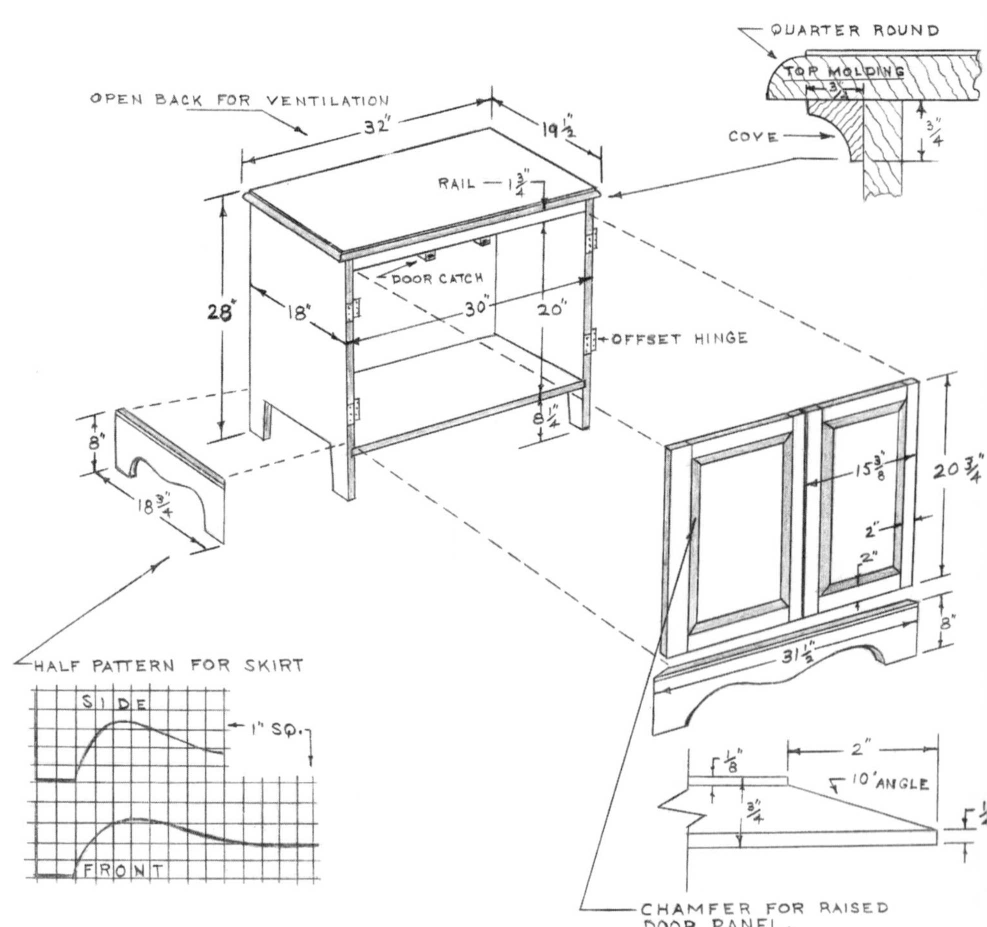

Fasten the top to the assembly by drilling countersunk pilot screw holes up through the cleats and into the underside of the top, then install #10, 1¼-inch flathead screws.

4. Use a router to make the quarter-round design around the three exposed edges of the top.

   Make or purchase ¾-inch cove molding. Cut the corners at a 45-degree miter joint, then fasten into place with #4 finishing nails.

5. Lay out and cut the three pieces to the skirt to the size and shape indicated in the drawing.

   Cut the corners at a 45-degree miter.

   Cut a chamfer or bevel along the top edge of the skirt.

   Secure into place by gluing and nailing with #6 finishing nails. Use a wet cloth to wipe away excess glue.

6. Construct the two raised-paneled doors to the overall size given in the drawing. (Refer to the front of the book for helpful hints on constructing raised-paneled doors.)

   Cut a ⅜-inch by ⅜-inch rabbet on the inner edge of the top, bottom, and hinged side of each panelled door. The cuts allow the doors to fit over the frame members.

   Hang the doors to the frame by using ⅜-inch offset hinges.

   Install the magnetic catches along with the doorpulls of your choice.

7. Set and fill all holes, then either remove hardware or cover with masking tape to sand the console to a smooth finished surface.

8. Select and apply the colored stain or paint of your choice. When dry, apply three to four coats of lacquer, rubbing between coats with fine steel wool or 400 emery paper. On the final coat rub with lemon oil and pumice. Preserve the finish with an occasional coat of paste wax or lemon oil. Re-install hardware.

## Project 52

# WEATHER CENTERS

### Materials Required

*Wood of Your Choice*

**1 board** 1″ × 5½″ × 14″
*or*
**1 board** 1″ × 5½″ × 10″
*or*
**1 board** 1″ × 5½″ × 5½″
**1 thermometer** 1″ × 2¾″ dia.
**1 barometer** 1″ × 2¾″ dia.
**1 hydrometer** 1″ × 2¾″ dia.

*Note:* The weather parts can be purchased at your local craft supply store.

### Procedures

1. Select the design of your choice, then, lay out and cut the board to the size indicated.
2. With a router and cove bit cut the cove design around the edge. A chamfered edge is equally attractive and can be cut with a hand plane.
3. Lay out and cut out the 2¼-inch-diameter instrument relief cuts. Make the cuts by first boring a ⅜-inch hole through the locations then inserting a sabre saw for cutting.
4. Sand the board to a smooth surface, then select and apply the stain or paint of your choice. When dry, apply several coats of clear lacquer finish, rubbing between coats with fine steel wool or pumice and lemon oil. Protect the finish by applying a coat of paste wax and buffing.

# Weather Centers

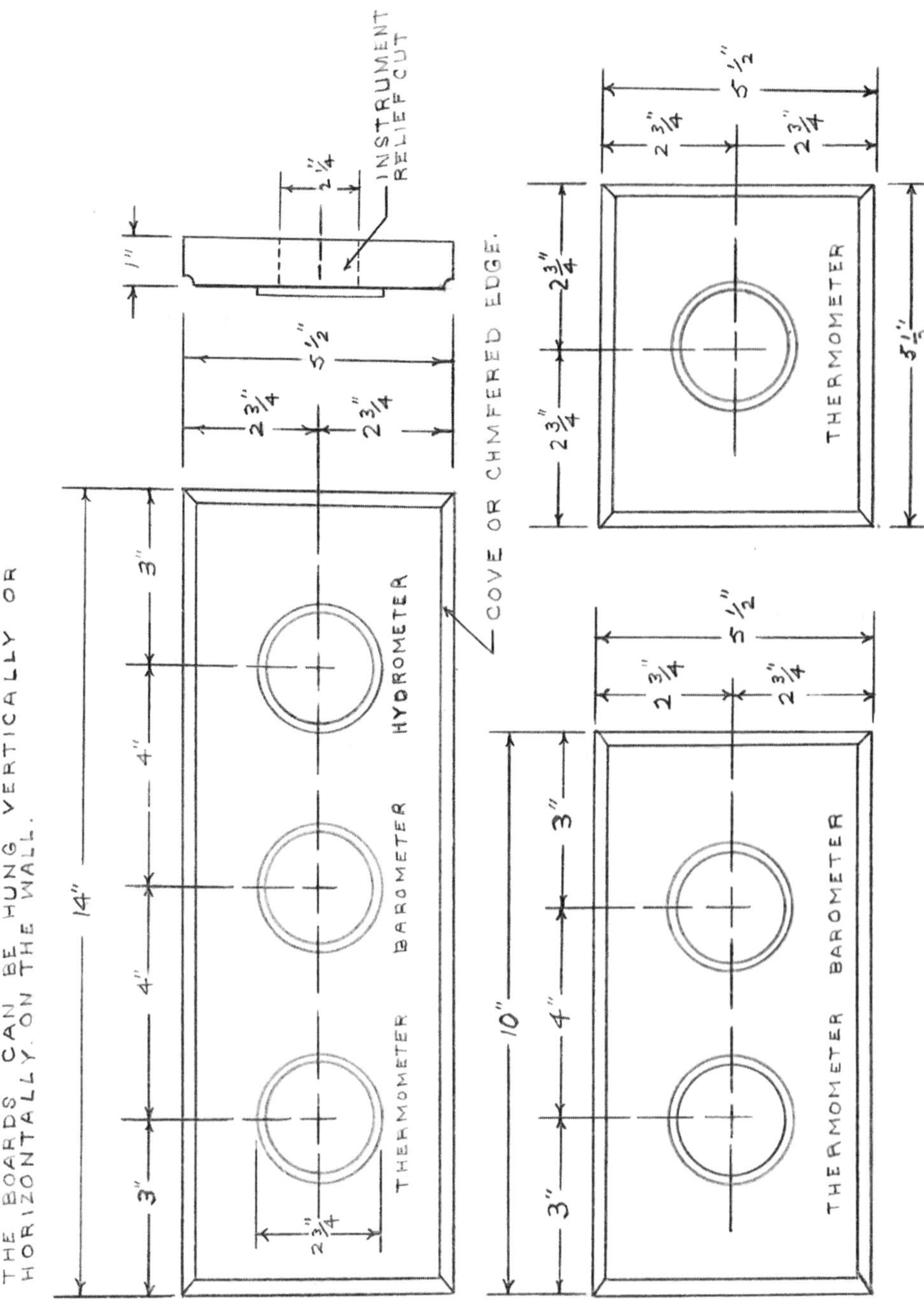

5. Insert the instrument clips into the 2¼-inch holes, then place the instruments in the clips.

## Project 53

# REFRESHMENT BAR

### Materials Required

*Wood of Your Choice*

**1 swing-open top** ¾″ × 13″ × 33″
**1 back section of top** ¾″ × 4″ × 33″
**2 sides** ¾″ × 16″ × 45¼″
**3 shelves** ¾″ × 15¾″ × 30½″
**1 back** ⅜″ × 31¼″ × 43¾″
**1 front swing-down panel** ¾″ × 12″ × 31¼″
**2 doors, raised-paneled** (overall size) ¾″ × 15⅝″ × 27½″
**2 skirts** ¾″ × 2¾″ × 32″
**6 offset hinges** ⅜″
**2 butt hinges**
**4 doorpulls**
**4 friction catches**
**2 drop-leaf lid supports**

### Procedures

1. Edge-glue enough stock to obtain the necessary widths for the top, shelves, sides, and doors.
2. Lay out and cut all of the stock to the size and designs shown in the drawing.
3. Make a ⅜-inch by ⅜-inch rabbet cut on the rear inner edge of the sides and bottom to receive the back.
4. Lay out the location of the shelves on the sides, then fasten the shelves to the sides using #6 finishing nails.

   *Note:* For added strength make dado joints into the sides to secure the shelves.

# Refreshment Bar

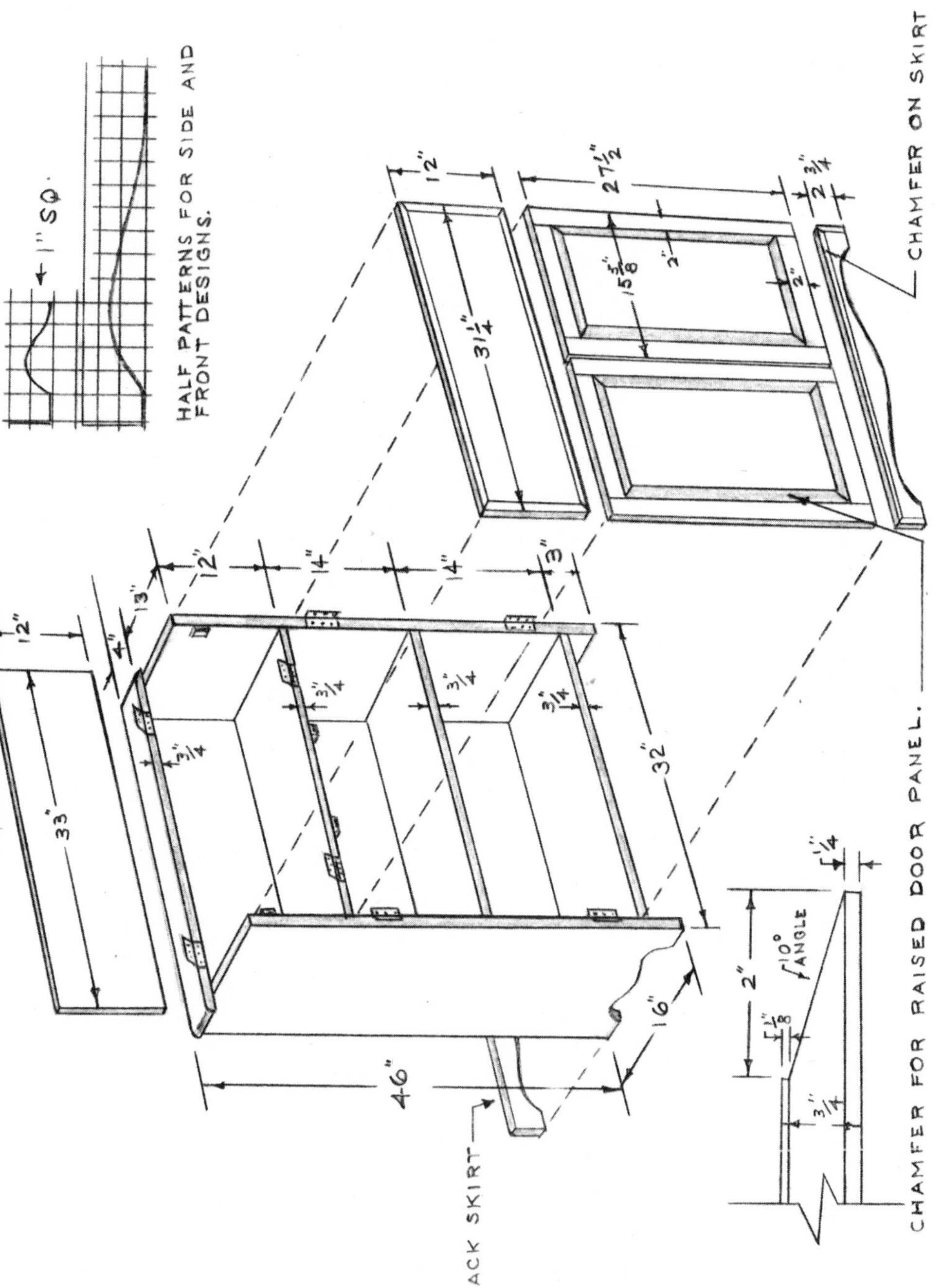

5. Fasten the back section of the top to the top edge of the sides by drilling counterbored pilot holes with #10, 1½-inch flathead screws. Make sure the piece overhangs the two ends and back by ½ inch.

   Secure a ¾-inch by 1-inch by 31¼-inch cleat to the underside of the top piece to fasten the back of the cabinet. Use glue and clamps to fasten.

6. Fasten the ⅜-inch back into the precut rabbets in the sides and bottom by using #4 finishing nails.

7. Cut a chamfer on two top pieces along with the swingdown door in the front.

8. Construct the two raised-paneled doors to the overall size shown in the drawing. Refer to the front of the book for instructions on making paneled doors.

   Cut a ⅜-inch by ⅜-inch rabbet completely around the inner edge of the top swingdown door and on three sides of the raised-paneled doors so that the doors will lap over the frame openings.

9. Install the doors to the frame by hanging them with ⅜-inch offset hinges. Install the swing-open top to the back stationary section of the top by using butt hinges.

10. Cut a chamfer on the three exposed edges of the front and back skirts, then secure them in place by gluing and nailing with #6 finishing nails.

11. Set and fill all holes, then sand the bar to a smooth surface. Use masking tape to cover the hardware, then stain or paint to the color of your choice. Apply four coats of polyurethane clear finish, rubbing between coats with lemon

oil and pumice. Preserve the finish with an occasional application of lemon oil.

12. Install the magnetic catches and doorpulls of your choice.

## Project 54

# WINE RACK

### Materials Required

*Wood of Your Choice*

**12 dowels** 1¼″ × 12″
**16 dowels** ⅜″ × 5¼″
**6 dowels** ⅜″ × 12″

### Procedures

1. Lay out and cut all of the dowels to the suggested size.
2. Construct the three horizontal levels first. Drill ⅜-inch-diameter holes halfway through the ends of the two 12-inch by 1¼-inch-diameter dowels and completely through the two center dowels. Be sure to use a spade bit and to make sure that the three levels are aligned perfectly. To avoid wood chipping when drilling, drill from one side until the bit point sticks out, then turn and complete the hole from the opposite side.
3. Insert the ⅜-inch by 12-inch dowels into the joints, using glue to fasten them. Be sure the levels are completely aligned.
4. Locate and drill thirty-two ⅜-inch-diameter holes to a depth of ½ inch to receive the vertical dowels. Drill into both sides of the middle level and into one side of the top level and one side of the bottom level.
5. Insert the eight vertical dowels into the bottom level, then continue to press the middle level onto these ⅜-inch dowels.

# Wine Rack

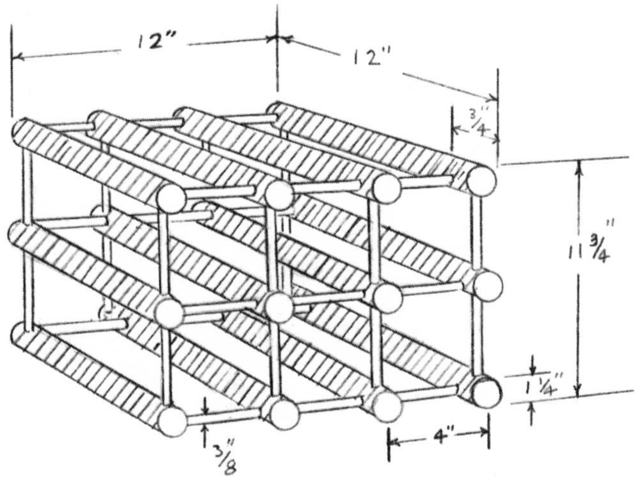


6. Insert the remaining dowels into the top of the middle level holes and continue in the same procedure as step five.

7. Before the glue dries, check with a framing square to see if the wine rack is aligned at right angles.

8. Sand the entire rack to a smooth surface.

9. Select and apply the paint or stain of your choice.

10. Apply several coats of oil or polyurethane, rubbing between coats with fine steel wool or pumice and oil.

## Project 55

# LOG BIN

### Materials Required

*Wood of Your Choice*

**1 top** 3/4″ × 8″ × 36″
**2 sides** 3/4″ × 18″ × 45 5/8″
**1 front** 3/4″ × 30″ × 34 1/2″
**1 back** 3/4″ × 46″ × 35 1/4″
**1 bottom** 3/4″ × 10 1/2″ × 34 1/2″
**2 drying rods** 3/4″ dia. × 36″

### Procedures

1. Rough cut stock to size, then use bar clamps to edge-glue the necessary pieces.

   *Note:* The back boards do not have to be glued together.

2. Lay out and cut the two sides to the shape and size indicated in the drawing. Cut a 3/8-inch by 3/4-inch rabbet the full length of the rear inner edge of both side pieces to receive the 3/4-inch by 46-inch by 35 1/4-inch back.

3. Lay out and cut the top, front, and bottom pieces to the size shown in the drawing.

   Cut 3/8-inch by 3/4-inch rabbets at each end of the top piece.

   Plane the stop chamfers on the front edge of the top and front pieces, then secure the top, back, front, and bottom to the sides using glue and #6 finishing nails or by drilling counter-bored pilot screw holes and installing #9, 1 1/2-inch flathead screws.

# Log Bin

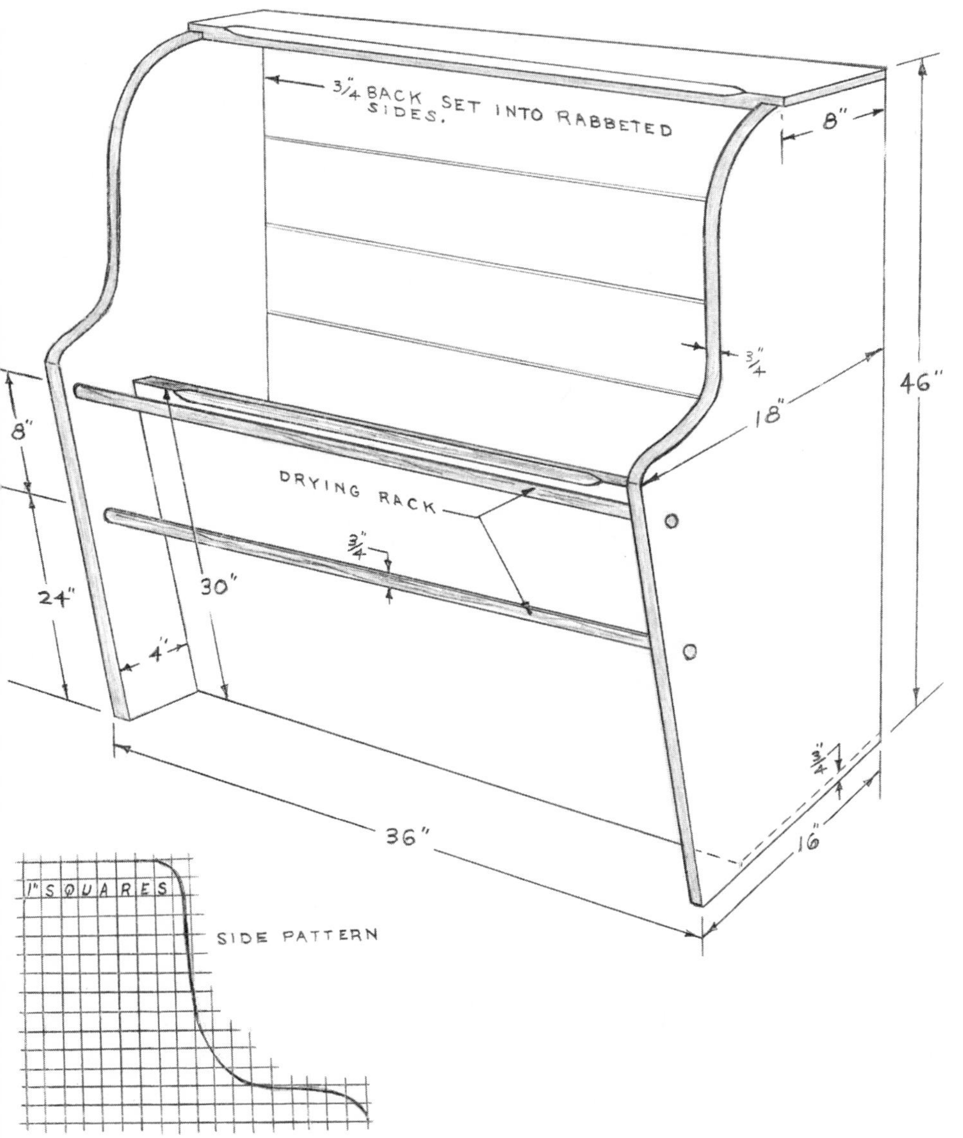
DRYING RACK POSITIONED
TO RECEIVE HEAT FROM A STOVE.
3/4 BACK SET INTO RABBETED SIDES.
8"
3/4"
18"
46"
8"
DRYING RACK
3/4"
24"
30"
4"
3/4"
36"
16"
1" SQUARES
SIDE PATTERN

4. Lay out and bore matching ¾-inch-diameter holes into the side pieces, then insert the ¾-inch-diameter by 36-inch dowels to complete the drying rack. Use glue to secure into place. Wipe off excess glue with a wet cloth.
5. Set and fill all holes. Plug screw holes, then scrape and sand the log bin to a smooth surface. Paint or stain to your color choice, then apply three coats of polyurethane clear finish, rubbing between coats with #400 emery paper. Preserve the finish with occasional coats of lemon oil.

## Project 56

# CONTEMPORARY KITCHEN TABLE AND CHAIR

### Materials Required

*Wood of Your Choice*

*Table:*

**4 legs** 3″ × 3″ × 27″
**2 long side rails** 3″ × 3″ × 48″
**2 end rails** 3″ × 3″ × 22″
**1 top** ¾″ × 30″ × 55″
**16 dowels** ½″ dia. × 5″
**plastic laminate** (optional)

*Chair:*

**2 front legs** 1¼″ × 2″ × 16½″
**2 back legs** 1¼″ × 2″ × 32″
**2 front and back rails** 1″ × 4″ × 14″
**2 side rails** 1″ × 4″ × 13″
**4 corner glue blocks** 2½″ × 2½″ × 3½″
**1 seat** ½″ × 13¾″ × 14¼″
**4 cleats** 1″ × 1″ × 10″
**1 foam rubber seat cushion pad** 1″ × 14½″ × 15″
**1 back support** ½″ × 5″ × 14″
**1 back foam rubber support cushion pad** 1″ × 5¾″ × 14¾″
**2 corner glue blocks for back support** 1¼″ × 1¼″ × 5″
**1 dowel** ⅜″ dia. × 36″
**contact cement**
**fabric tape**
**1 square yard (approximately) of the fabric of your choice.**

# Contemporary Kitchen Table and Chair

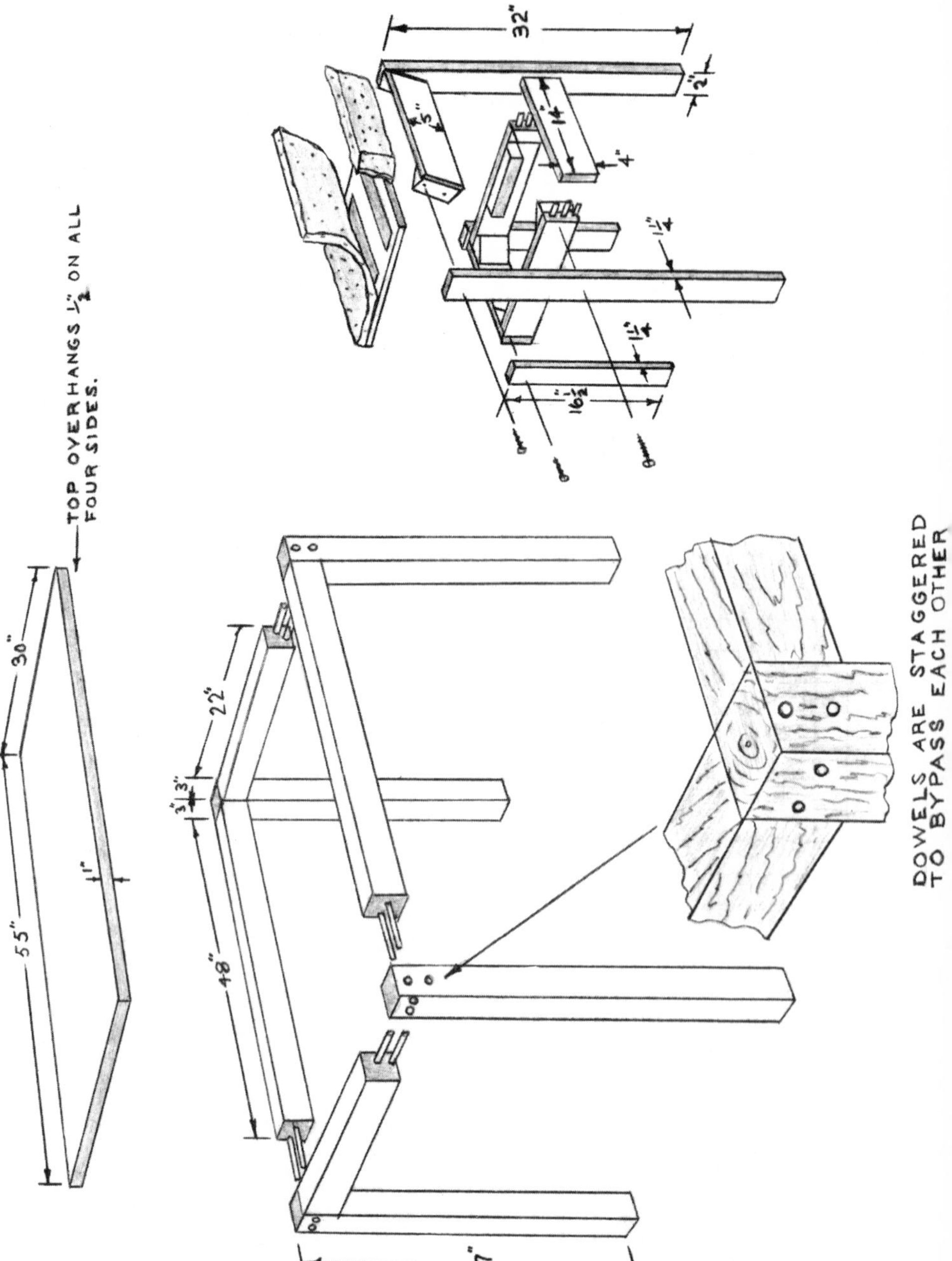

## Procedures

### Chair

1. Lay out and cut all of the chair parts to the sizes and shapes indicated in the list of materials and drawing.

2. Lay out and drill ⅜-inch-diameter dowel holes through the front and back rails along with the side rails. For accuracy use a doweling jig to align the holes. Secure the frame together by applying glue to the dowel joints.

3. Cut the 2½-inch by 3½-inch diagonal corner glue blocks to reinforce the corners, then fasten them into place with glue and #4 finishing nails.

   Nail and glue 1-inch by 1-inch by 10-inch cleats flush to the inner top edge of all four rails.

4. Drill pilot screw holes up through the cleats and into the ½-inch seat panel.

   Prefasten the seat panel to the cleats with #9, 1¼-inch screws. Disassemble seat until foam and fabric is applied.

5. Drill counterbored pilot screw holes through each leg and into the side rails of the seat frame, then secure the legs to the rails with #10, 1½-inch flathead screws. Cover counterbored holes with hardwood plugs.

6. Nail and glue the corner blocks flush to the ends of the back support board.

   Hold the foam rubber back in place using tape and staples. Use ⅜-inch staples to fasten the fabric to the back support bottom edge and glue blocks.

Drill counterbored pilot screw holes through the top of the legs and into the back support glue blocks to secure the back support flush to the top of the legs at approximately a 70-degree angle.

Use two #10, 1½-inch flathead screws covered with hardwood plugs.

*Note:* If a staple gun is not available, use furniture tacks to fasten the fabric to the back.

7. Sew the seat slipcover together, leaving the rear edge open.

   Tape or glue the 1-inch foam rubber to the seat, then slip the cover completely over the foam seat. Sew up the open rear fabric.

   Locate the position of each predrilled pilot hole in the bottom side of the seat, then with a sharp, stiff knife cut a small section of fabric away to allow screws to enter freely.

   Secure the upholstered seat to the predrilled cleats with #9, 1¼-inch flathead screws.

8. Cover all fabric, then apply the stain or paint of your choice. When dry, apply several coats of clear polyurethane finish, rubbing between coats with fine steel wool or #400 emery paper. Preserve the finish with an occasional coat of paste wax or lemon oil.

## Table

1. If the table top is to be made from solid wood, it will have to be edge-glued and clamped with bar clamps. If it is to be covered with plastic laminate, then ¾-inch-thick particle board or plywood can be used.
2. Cut all of the pieces to the size shown in the drawing. It may be necessary to laminate the 3-inch by 3-inch by 27-inch legs.

3. Start by securing the legs to the end rails.

   Drill two ½-inch-diameter dowel holes completely through the 3-inch legs and 2 inches into the rails. Center the holes to the width of the leg, then space them ¾ inch and 2 inches from the top of the leg.

   Fasten the long side rails to the legs by drilling two ½-inch-diameter holes completely through each leg and 2 inches into rails. Both holes should be located approximately 1⅜ inches from the top of the leg and ¾ inch from the sides. Use glue when joining doweled corners.

4. Position the top, then drill counterbored pilot screw holes to secure the top to the rails with #10, 2½-inch flathead screws.

5. Finish the table in the same manner the chair was finished. If the top is not painted or given a clear finish, cover it with plastic laminate.

   *Note:* The top can be covered with laminate at your local lumber mill shop.

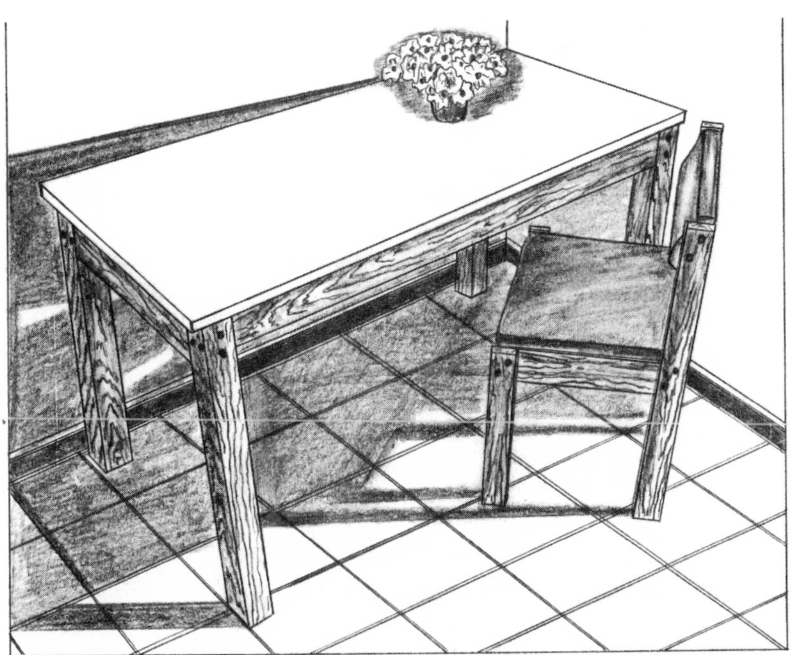

## Project 57

# SERVING BOARD

### Materials Required

*Wood of Your Choice*

**16 strips** ½″ × 1⅛″ × 10″
**waterproof glue**

*Note:* The pieces are cut to a 10-inch length to allow for waste due to the saw kerf.

### Procedures

1. Lay out and cut sixteen ½-inch by 1⅛-inch by 8-inch strips. Eight of the strips should be mahogany or walnut and eight of the strips should be maple or birch.
2. Glue and clamp the sixteen alternate strips together, making sure that all of the glue joints are clamped tight.
3. Scrape away all traces of glue, then plane the top and bottom surfaces smooth.
4. Set the saw for a ½-inch cut, then make fifteen cuts across the glued-up board. (Refer to drawing.)
5. Glue and clamp the fifteen ½-inch by 8-inch strips together, making sure to shift one space for each strip. (Refer to drawing.)
6. Cut off the protruding ends. (Refer to drawing.)
7. Scrape off all traces of glue, then plane both surfaces even.
8. Sand the board to a smooth surface, then apply several coats of finishing oil. Rub between each coat with fine steel wool.

# Serving Board

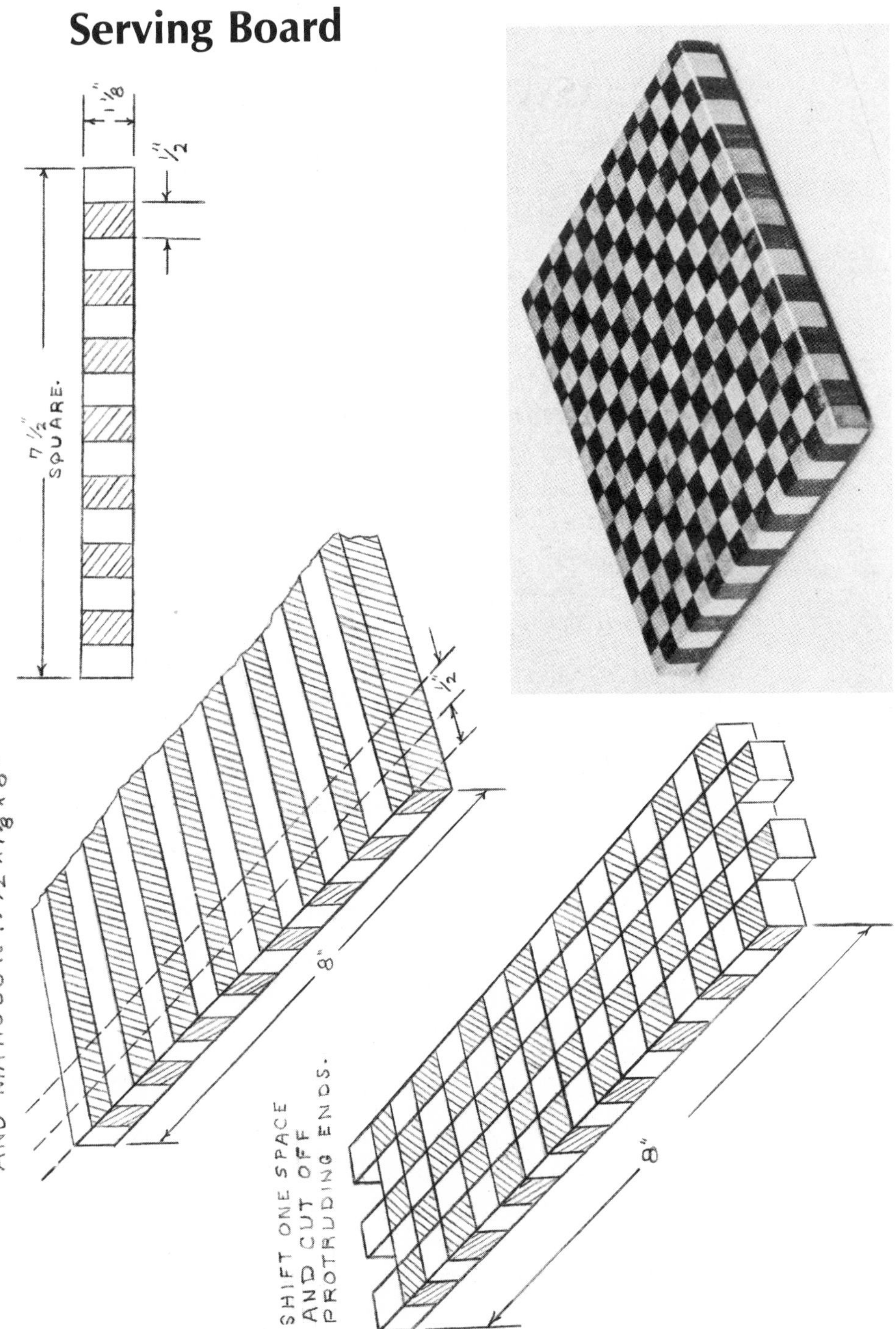

## Project 58

# DELUXE STEREO CABINET

### Materials Required

*Wood of Your Choice*

**2 sides** ¾″ × 22″ × 37¼″
**1 top (A)** ¾″ × 18″ × 22″
**1 top (B) dust cover** ¾″ × 18″ × 20″
**1 top (C)** ¾″ × 18″ × 20″
**1 front to dust cover** ¾″ × 8½″ × 20″
**2 dividers to top section** ¾″ × 8½″ × 17¾″
**1 back to top section** ¼″ × 7¾″ × 59¼″
**1 upper shelf** ¾″ × 22″ × 60½″
**1 lower shelf** ¾″ × 17½″ × 60½″
**2 dividers to lower section** ¾″ × 17½″ × 24″
**1 baseboard** ¾″ × 3¼″ × 60½″

### Procedures

1. Determine if you are to use solid core birch, plywood, or cherry hardwood. If you do not use plywood, you will have to use bar clamps to edge-glue all of the rough cut stock to the required width for the necessary pieces.
2. Lay out and cut all of the pieces to the design and shape shown in the drawing. Be sure all pieces are cut square (90 degrees).
3. On the top rear edge of the side pieces make a ⅜-inch by ⅜-inch by 8-inch stop rabbet cut. On the rear inner edge of the three top pieces and along the top shelf make a ⅜-inch by ⅜-inch rabbet cut. All of these cuts are to receive the perforated hardboard back.
4. Scrape and sand all pieces in preparation for assembly.

# Deluxe Stereo Cabinet

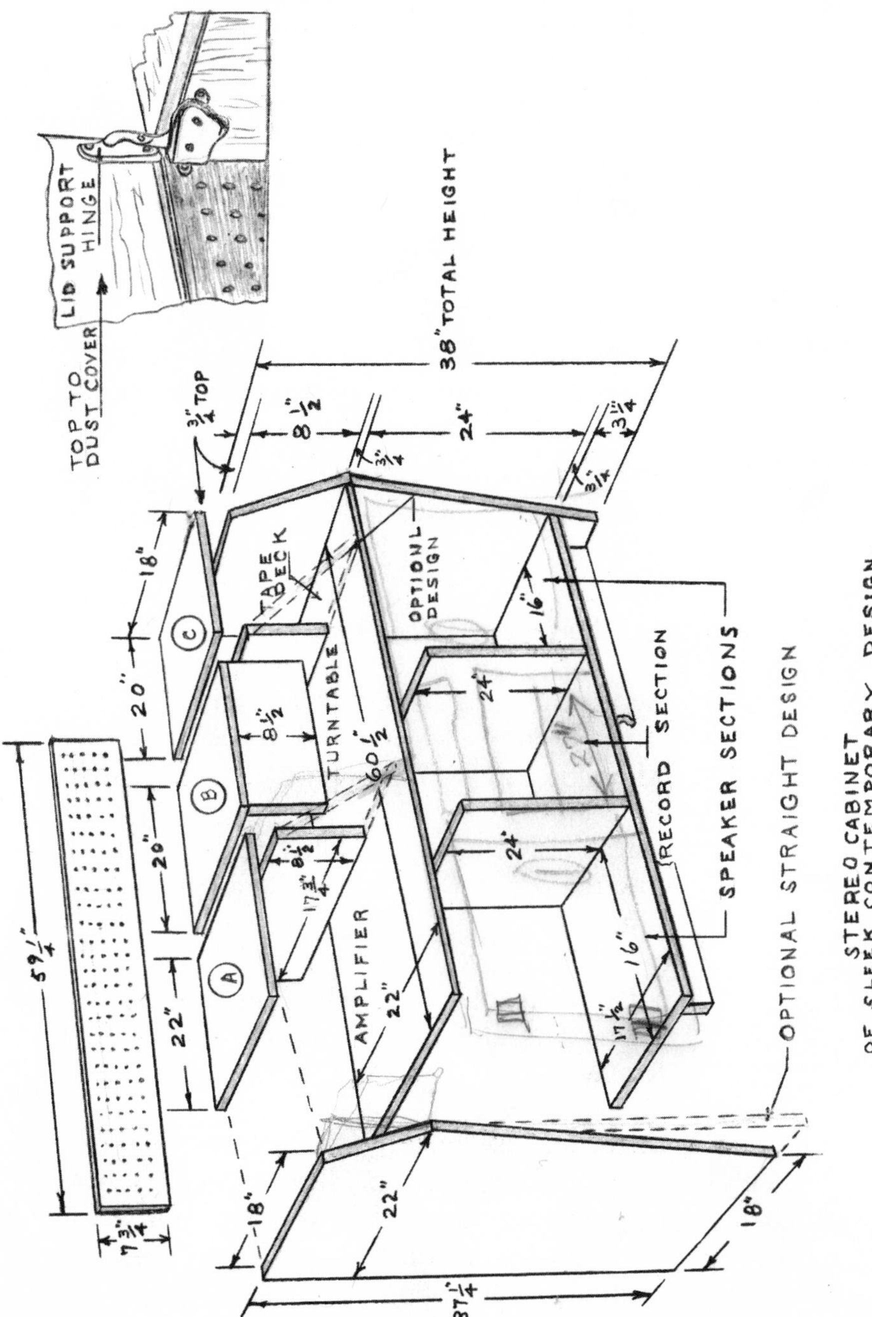

5. Lay out and cut matching dadoes into the side pieces, or drill counterbored pilot screw holes instead. Fasten the two sides to the shelves with #10, 1¾-inch flatheads.

6. Install the bottom dividers into place with #6 finishing nails. Fasten top (A) and top (C) to the top edge of the sides and the dividers. Use glue and #6 finishing nails.

7. Construct the dust cover by recessing the front piece approximately ½ inch from the front edge of the top. The overhang will allow a finger grip for the user.

   Position and install two lid support hinges into place.

   *Note:* The rear inner edge of the dust cover top will have to be chamfered in order to clear the hardboard back when opening

8. Recess the baseboard approximately 2 inches from the front edge of the bottom shelf. Fasten with glue and #6 finishing nails.

9. Install the ¼-inch perforated hardboard back into the precut rabbets using ⅞-inch flathead brads.

10. Set and fill nail holes, plug counterbored screw holes, than sand the cabinet to a smooth surface.

11. Select and apply the stain or paint of your choice. When dry, apply three coats of polyurethane clear finish. Rub between coats with #320 emery paper or fine steel wool. Preserve the finish with an occasional coat of lemon oil.

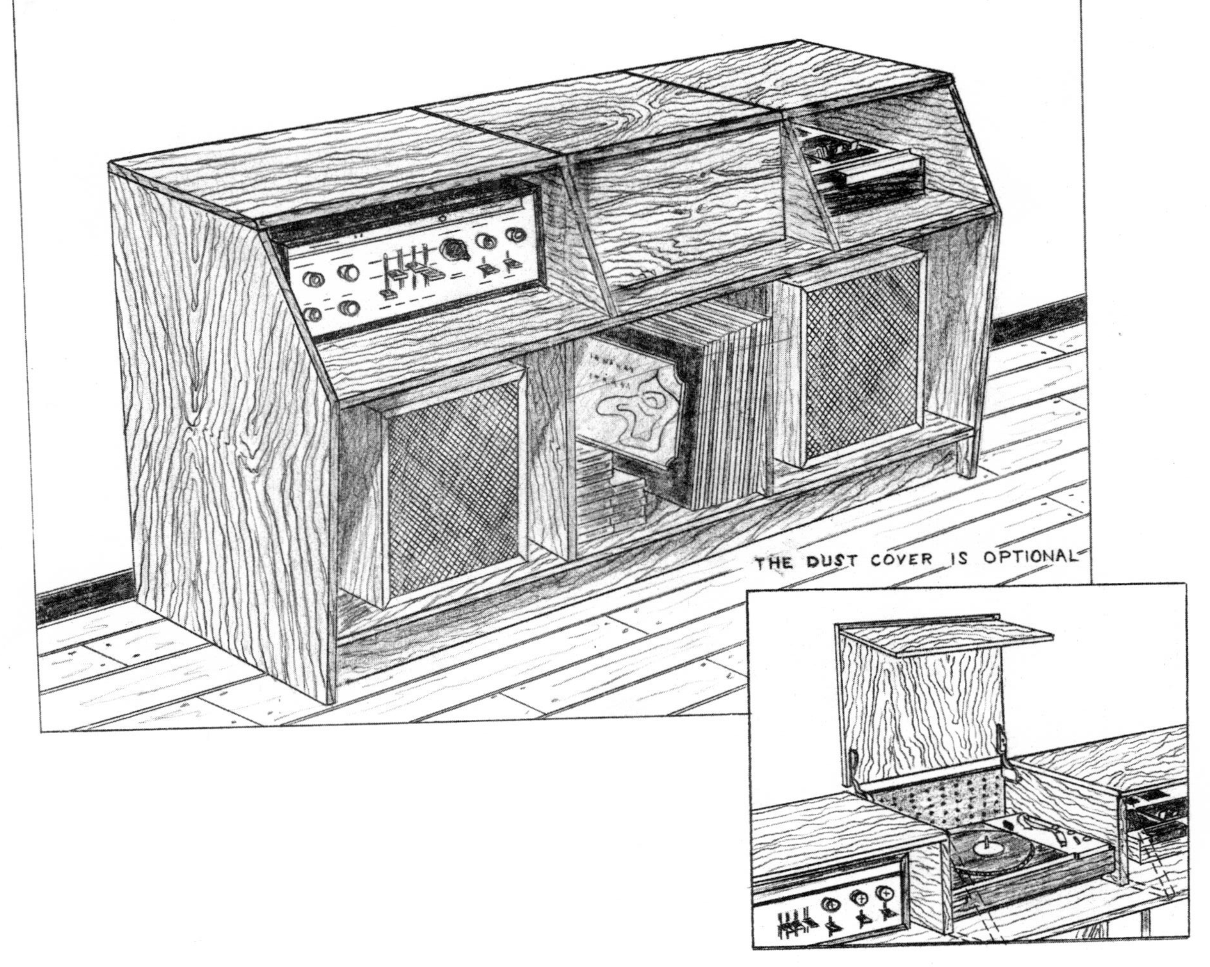
THE DUST COVER IS OPTIONAL

## Project 59

# CORNER CUPBOARD

### Materials Required

*Wood of Your Choice*

**5 shelves** ¾″ × 22″ × 32″
**1 top** ¾″ × 22″ × 32″
**2 sides** ½″ × 17″ × 78″
**2 stiles** ¾″ × 4½″ × 78″
**1 top rail** ¾″ × 12″ × 24″
**1 middle rail** ¾″ × 3″ × 24″
**1 base skirt** ¾″ × 4½″ × 24″
**2 frame moldings** 1¼″ half-round × 78″
**1 frame molding** 1¼″ × 30″
**2 doors, raised-paneled (overall size)** ¾″ × 11¹⁵⁄₁₆″ × 22⅜″
**2 magnetic catches**
**2 pairs of hinges of your choice**

### Procedures

1. Edge-glue enough stock to meet the required widths for the top, shelves, and doors.
2. Lay out and cut the two sides along with the top and shelves. Make all side cuts at a 45-degree angle.
3. Lay out and cut the two stiles to the size shown in the drawing.

   Cut a full-length rabbet into the outer edge of the stiles so that the ½-inch back will set in at a 45-degree angle.
4. Lay out and cut the two sides to the size shown in the drawing.

   Nail the two sides into the precut rabbets in the stiles.

# Corner Cupboard

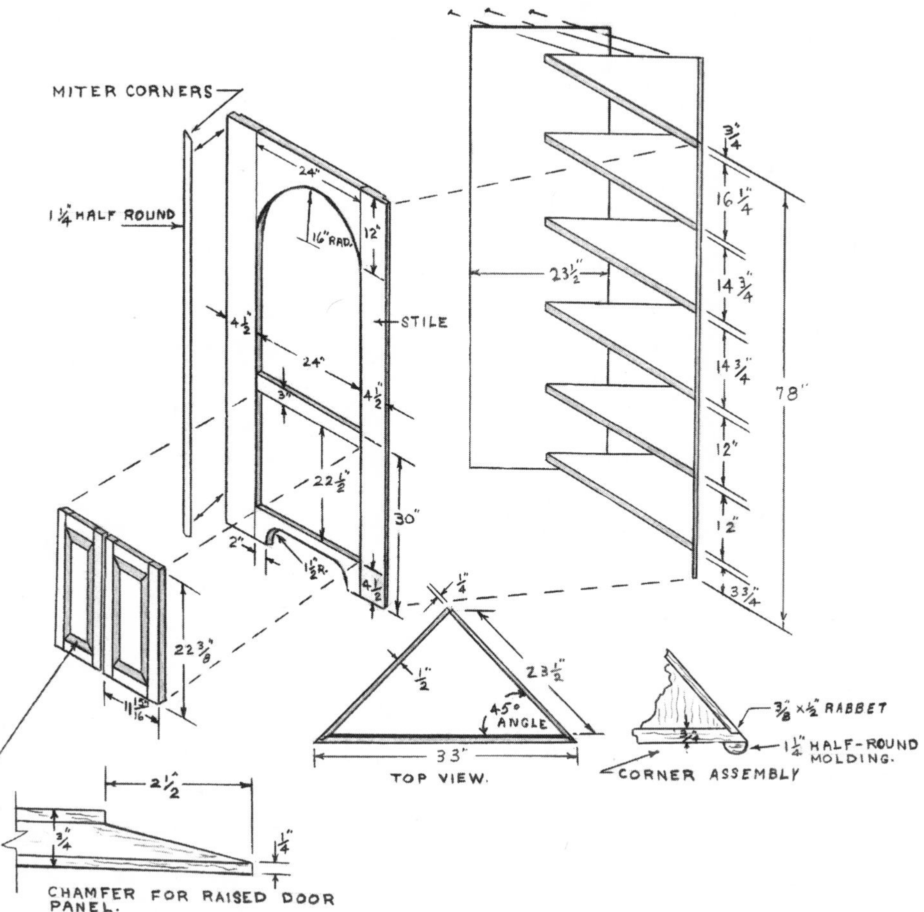


Lay out the proper spacing for the shelves, then nail the shelves to the sides and stiles using #6 finishing nails.

5. Lay out and cut the two rails and skirt to the size and design shown. Fasten these pieces in between the stiles and to the front edge of the shelves and top. Glue blocks should be attached to the sides and beneath the top for added fastening surface.

6. Lay out and cut the mitered corners on the 1¼-inch half-round molding, then nail the three pieces flush to the outside edge of the cupboard. Use #4 finishing nails.
7. Construct the raised-paneled doors to the overall dimensions given.

   *Note:* Refer to the front of the book for instructions.
8. Install either surface or butt hinges to hang the two doors, then install two magnetic catches along with two doorpulls. The doors are made flush to the frame.
9. Set and fill all holes, then sand the entire cupboard to a smooth surface. Use masking tape to cover the hardware, then stain or paint to the color of your choice. Apply three coats of polyurethane finish, rubbing between coats with fine steel wool or 230 emery paper. Preserve the finish with an occasional application of lemon oil.

## Project 60

# BOOK ENDS

### Materials Required

*Wood of Your Choice*

*For One Pair:*
**2 disks** ¾″ × 6½″ × 6½″
**2 disks** ¾″ × 5½″ × 5½″
**2 bases** ¾″ × 2½″ × 6½″
**2 book rests** ⅛″ × 4″ × 4″
**4 wood screws** #3, ⅜″ flathead
**4 wood screws** #9, 1½″ flathead

### Procedures

1. Lay out and cut all pieces to the suggested size and shape indicated in the drawing.
2. With a chisel cut a ⅜-inch by ¾-inch by 4-inch gain under the base piece to receive the sheet metal plate.
3. Lay out and cut a ¼-inch chamfer around the top of the base piece.
4. Fasten the two disk pieces together by gluing and clamping.
5. Locate and drill four countersunk pilot holes through the base and into the round pieces, then attach the base to the two disk pieces with #9, 1½-inch flathead wood screws.
6. Locate and drill two countersunk pilot holes through the ⅜-inch sheet metal and into the base piece, then fasten the sheet metal book rest to the base with #3, ⅜-inch flathead screws.

# Book Ends

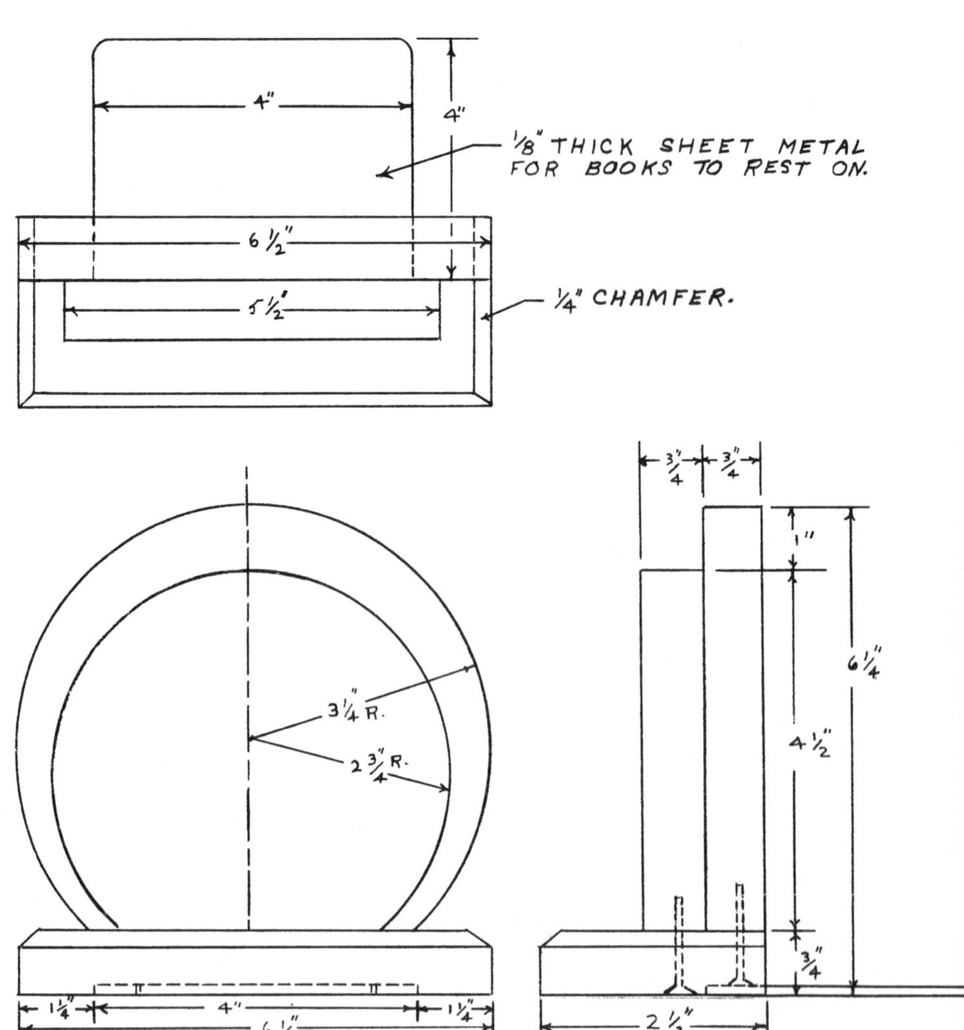
4"
4"
1/8" THICK SHEET METAL FOR BOOKS TO REST ON.
6 1/2"
5 1/2"
1/4" CHAMFER.
3/4
3/4
1"
6 1/4
3 1/4 R.
2 3/4 R.
4 1/2
3/4
1 1/4"
4"
1 1/4"
6 1/2
2 1/2"

7. Scrape off all traces of glue, then file and sand the project to a smooth surface.
8. Apply several coats of clear lacquer finish, rubbing between coats with fine steel wool or pumice and lemon oil. Protect the finish by applying a coat of paste wax and buffing.

## Project 61

# SEWING CENTER

### Materials Required

*Wood of Your Choice*

**2 sides** ¾″ × 15¼″ × 48″
**1 top** ¾″ × 15¼″ × 38½″
**1 bottom** ¾″ × 15¼″ × 38½″
**2 shelves** ¾″ × 14⅝″ × 38½″
**1 back** ⅝″ × 39¼″ × 47⅜″
**2 stiles** ¾″ × 2″ × 47¾″
**1 top rail** ¾″ × 2″ × 36″
**1 base rail** ¾″ × 3¼″ × 36″
**3 storage dividers** ¾″ × 9¼″ × 10″
**1 storage divider** ¾″ × 10″ × 21″
**2 dowels** ¼″ dia. × 36″
**8 continuous hinges**
**2 casters**
**3 friction catches**

### Procedures

1. Begin by deciding if this project is to be made from plywood or standard lumber. If you choose the standard lumber, use bar clamps to edge-glue the stock to the required widths.
2. Lay out and cut all of the stock to the size specified in the drawing.
3. Make a ⅜-inch by ⅝-inch rabbet cut on the rear inner edge of the top, bottom, and two sides. The rabbet cuts should stop approximately ⅝ inch from the top edge of the side pieces. These cuts are to receive the ⅝-inch back.

# Sewing Center

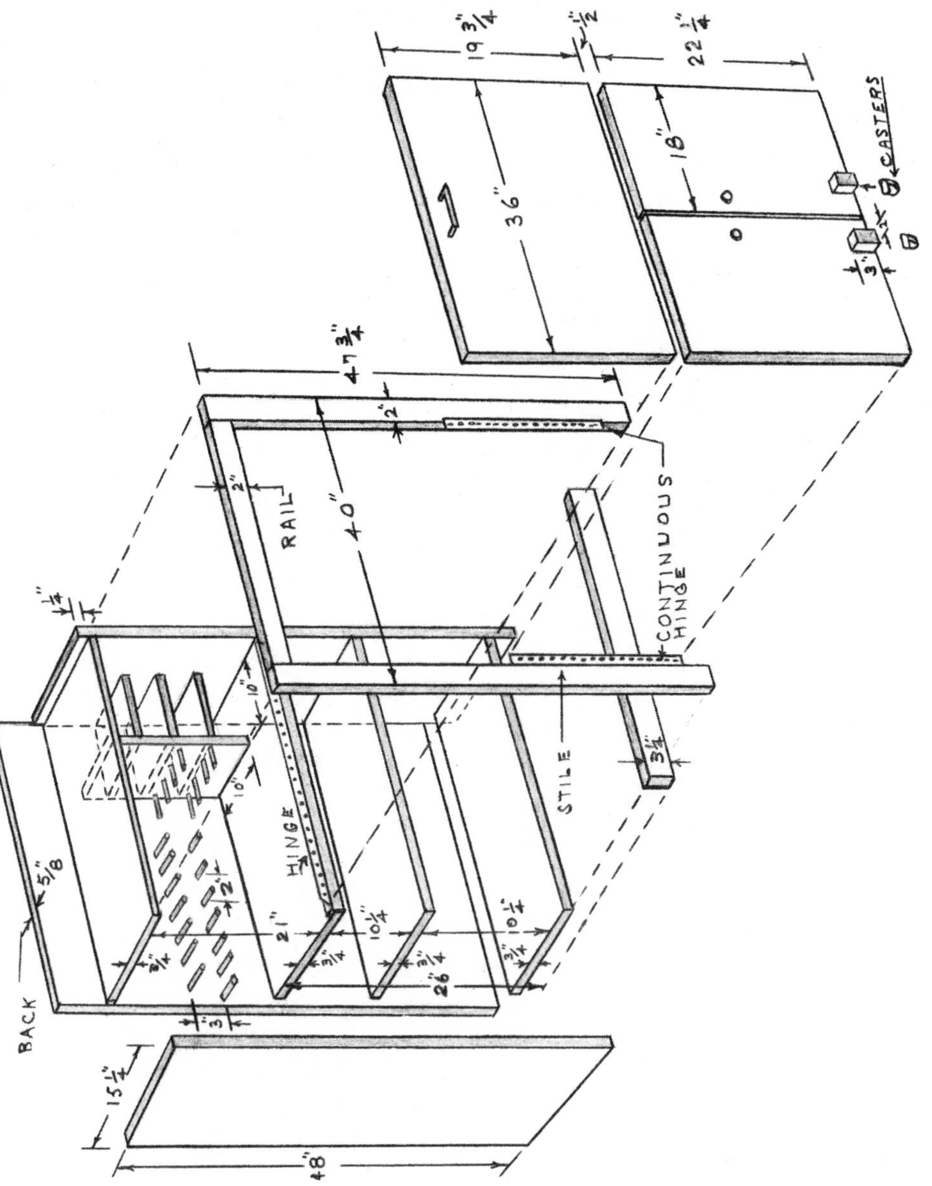

4. Lay out the position of joints, then assemble the sides, top, bottom, and back using #6 finishing nails according to the drawing. Allow the top to be recessed below the side pieces approximately ¼ inch.

5. Lay out and nail into place the shelves and shelf storage dividers using #6 finishing nails.

6. Fasten the stiles and rails onto the cabinet front to complete the face frame. Use glue and nails.

7. Lay out and drill twenty-two ¼-inch-diameter holes for the dowels to hold the spools of thread in the storage compartment.

   Cut the twenty-two ¼-inch-diameter by 2½-inch dowels and secure them into the holes with glue.

8. Fasten the continuous hinges into place, as shown in the drawing, then secure the casters by screwing them to the blocks mounted on the lower doors.

   *Note:* To prevent the doors from swinging out wider than the pull-down workshelf, attach a small chain with eye hooks to the door and cabinet frame. Install the doorpulls of your choice.

9. Use masking tape to cover the hardware, then set and fill all holes. Sand the cabinet to a smooth surface, then apply the stain or paint of your choice. When dry, apply three coats of clear lacquer, rubbing between coats with fine steel wool or #320 emery paper. Preserve the finish with an occasional coat of lemon oil.

## Project 62

# PORTABLE TELEVISION STAND

### Materials Required

*Wood of Your Choice*

**1 top** ¾″ × 15″ × 30½″
**1 bottom** ¾″ × 15″ × 30½″
**2 sides** ¾″ × 15½″ × 24″
**1 back** ¼″ × 18¾″ × 31¼″
**1 design rail** ¾″ × 4″ × 30½″
**1 skirt** ¾″ × 2″ × 30½″
**4 casters**

### Procedures

1. Rough cut the stock for the top, bottom, and two sides, then use bar clamps to edge-glue the pieces to the necessary widths.
2. Lay out and cut the top, bottom, and two sides to the size and shape shown in the drawing.
3. Make a rabbet cut on the rear inner edge of the top, bottom, and two sides to receive the back.
4. Fasten the sides to the two shelves by drilling counterbored pilot holes and securing with #9, 1½-inch flathead screws.

   Fasten the back into the precut rabbets with ⅞-inch flathead brads.

   *Note:* Before the back is fastened permanently, check the assembly for squareness (90 degrees).
5. Lay out and cut the top rail and base skirt to the size and design shown in the drawing. Fasten these two pieces to the assembly using glue and #6 finishing nails.

# Portable Television Stand

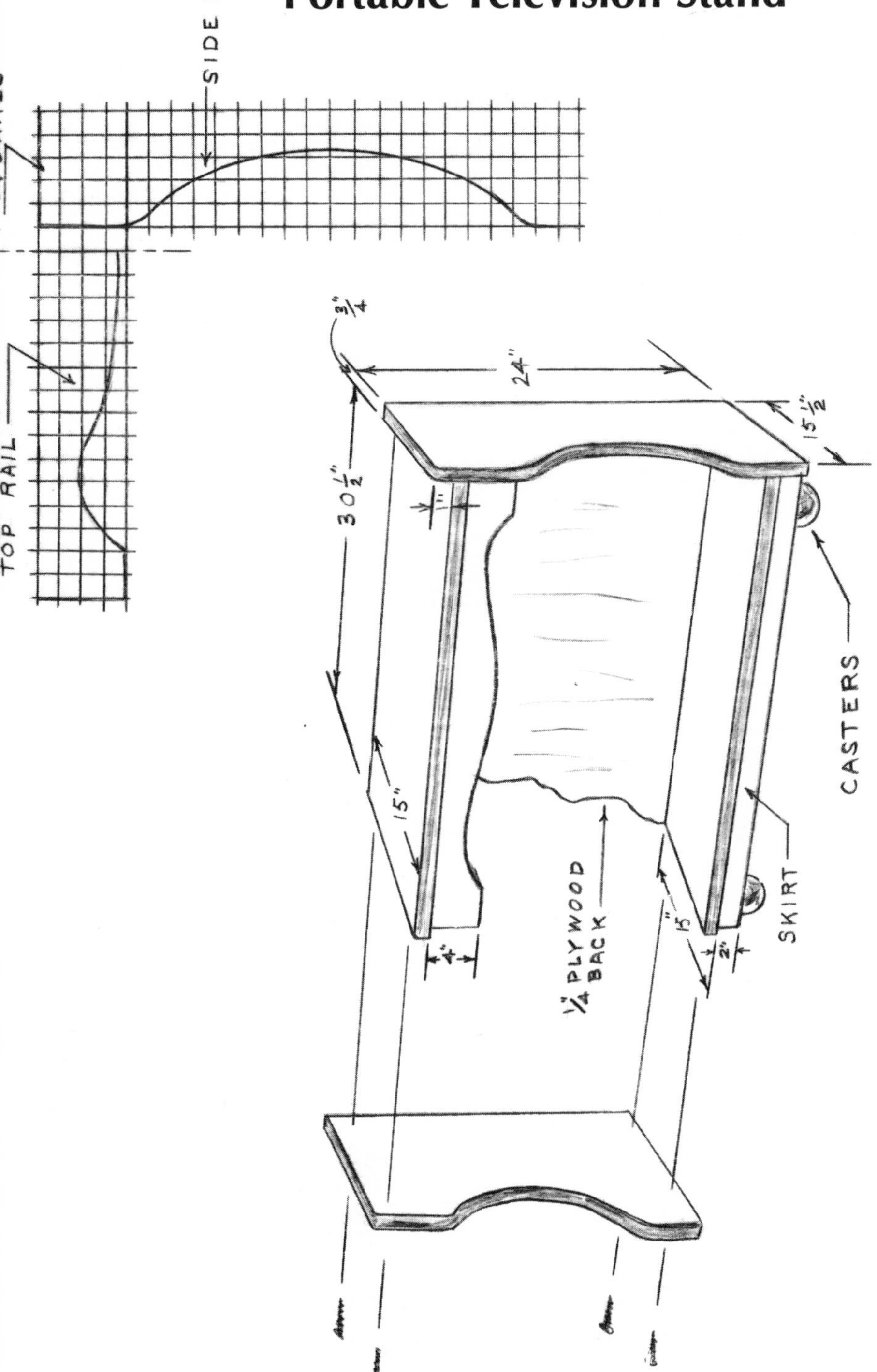

6. Purchase four casters of your choice, then secure them to the underface of the bottom shelf.
7. Plug all counterbored holes, set and fill all nail holes, then sand the stand to a smooth finished surface. Select and apply the stain or paint of your choice. When dry, apply three coats of polyurethane clear finish, rubbing between coats with fine steel wool or #320 emery paper. Preserve the finish with an occasional coat of lemon oil.

## Project 63

# TRESTLE COFFEE TABLE

### Materials Required

*Wood of Your Choice*

**1 top** 1″ × 18″ × 54″
**2 top cleats** 1½″ × 2″ × 11″
**2 legs** 1½″ × 6″ × 11″
**2 leg bases** 2″ × 2″ × 11″
**1 center rail** 2″ × 2½″ × 46″
**2 hardwood pins** ½″ dia. × 3″

### Procedures

1. Using bar clamps, edge-glue up enough stock to meet the required width for the top and two legs.

2. Select the design of your choice, then lay out and cut the legs to shape.

   Lay out and cut the 1-inch by 2½-inch mortise into the center trestle legs to receive the center rail.

   Fasten the top cleats along with the base to the center trestle leg by using #12, 3-inch flathead wood screws recessed into counterbored pilot holes.

   *Note:* If the tenon fits too tight, adjust it by filing it slightly. Lay out and bore a ½-inch-diameter hole where the tenon clears the outside edge of the leg. Fasten the ½-inch-diameter by 3-inch pins through the prebored holes in the center rail tenons.

# Trestle Coffee Table

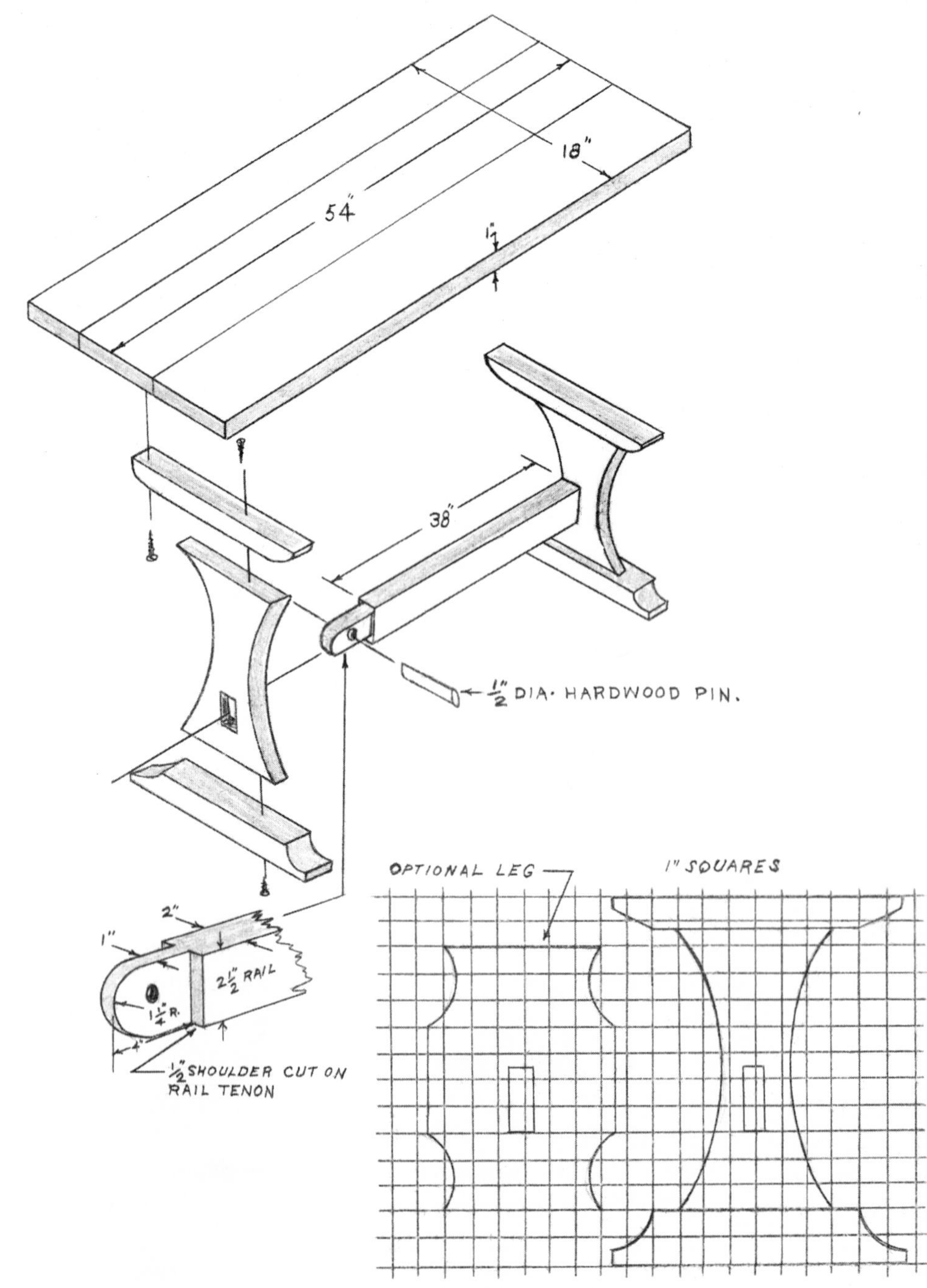

4. Cut the top to the size indicated in the drawing. The edge is attractive if it is rounded over to a quarter-round shape; use a router for this process.

   Secure the top to the trestle leg cleats by drilling countersunk pilot holes up through the cleats and into the underface of the top. Use #10, 2½-inch screws.

5. Sand the table to a smooth finished surface, then select and apply the stain of your choice. When dry, apply four coats of clear finishing oil, rubbing between coats with pumice and lemon oil. Preserve the finish with an occasional coat of lemon oil.

   *Note:* It is easier to disassemble the table before sanding and finishing.

## Project 64

# LAZY SUSAN

### Materials Required

*Wood of Your Choice*

**1 round tray** ⅝″ × 16″ dia.
**1 base** 1″ × 8½″ × 8½″
**1 bearing mechanism** 4″ × 4″
**8 wood screws** #6, ½″ flathead

*Note:* The bearing mechanism can be purchased at a local hardware store.

### Procedures

1. Edge-glue the stock to reach the recommended width.
2. Lay out and cut all stock to the suggested size and design.
3. The edge of the lazy susan is equally attractive plain or with a cove cut routed around.
4. Fasten the lazy susan bearing mechanism to the center of the base and tray bottom. Use #6, ½-inch flathead screws.
5. Sand the entire project to a smooth surface. Select and apply a stain of your choice. When thoroughly dry, apply several coats of finishing oil, rubbing lightly between each coat with fine steel wool or pumice and oil.

# Lazy Susan

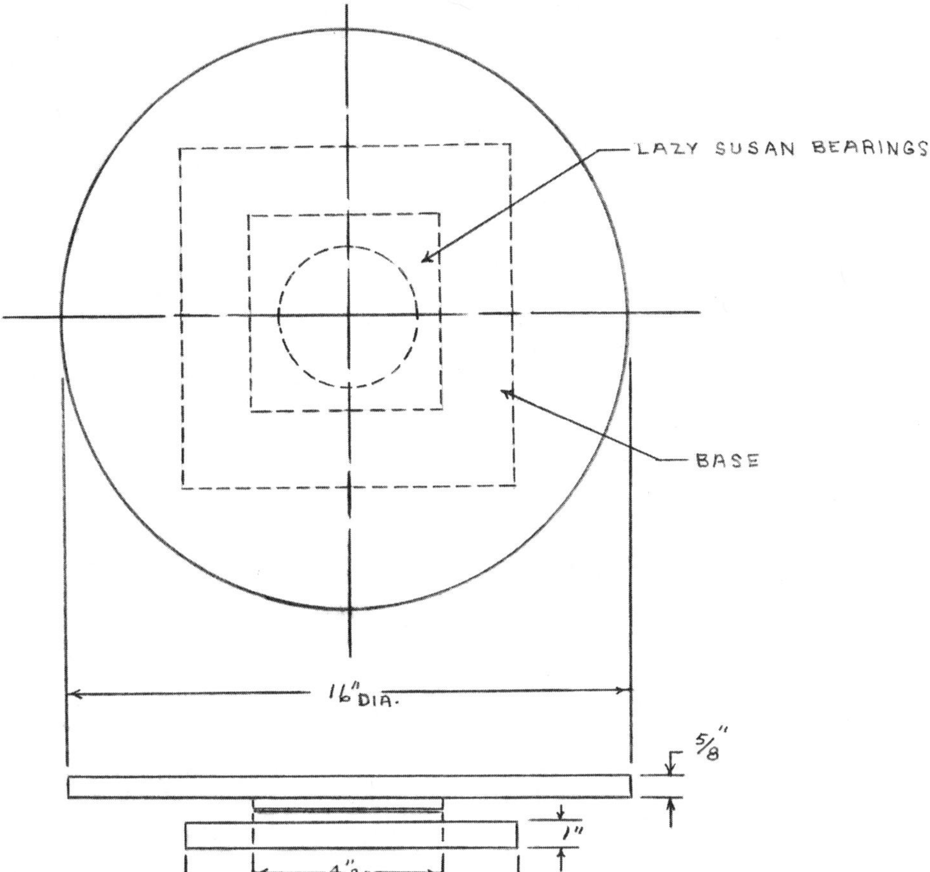

LAZY SUSAN BEARINGS

BASE

16" DIA.

5/8"

1"

4" SQ.

8 1/2 SQ.

## Project 65

# DINING TABLE

### Materials Required

*Wood of Your Choice*

**1 top** 1¼″ × 34″ × 72″
**2 legs** 3″ × 9″ × 21¾″
**2 leg top cleats** 3″ × 4″ × 28″
**2 leg bases** 4″ × 4″ × 28″
**1 center rail** 2″ × 5″ × 62″
**2 hardwood pins** ½″ dia. × 3″

### Procedures

1. Cut stock then use bar clamps to edge-glue up enough stock for the top and legs.

2. Lay out and cut the legs to the size and shape shown. Cut the 1-inch by 5-inch mortise into the center of the legs to receive the center rail.

   Fasten the top leg cleats along with the base to the center trestle legs by using counterbored pilot holes and the screws shown in the drawing.

3. Lay out and cut the center rail to the size indicated in the drawing. Make ½-inch by 5-inch by 6-inch shoulder cuts to complete the tenon at each end of the center rail. The tenon should fit snugly into the mortises in the legs.

   *Note:* If the tenon fits too tight, make the adjustment by filing it slightly. Lay out and bore two ½-inch-diameter by 3-inch pins through the bored holes in the tenon.

# Dining Table

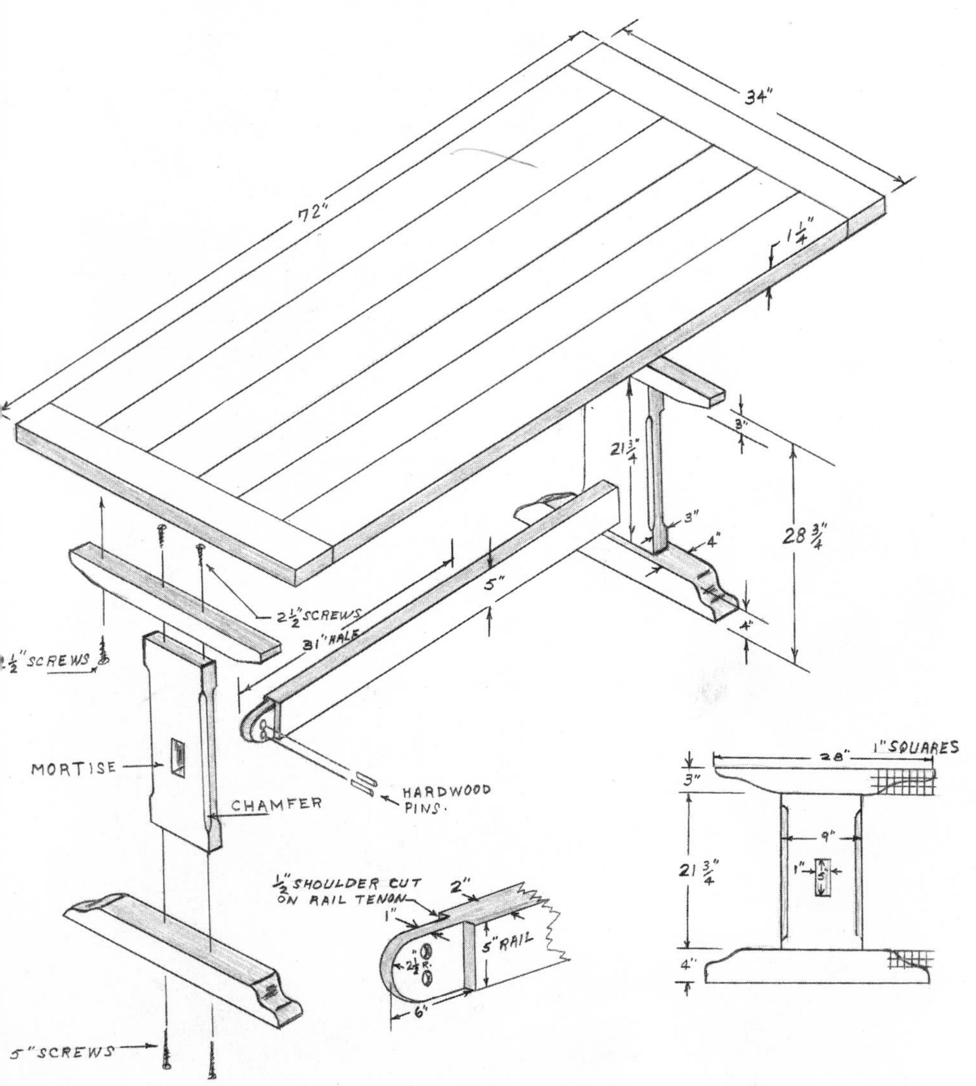

4. Lay out and cut the top to the correct size. The edge is attractive if it is rounded over to a quarter-round design. Secure the top to the cleats by drilling countersunk pilot holes up through the cleats and into the underface of the top. Fasten with the screws shown in the drawing.
5. Sand the table to a smooth finished surface, then select and apply the stain of your choice. When dry, apply four coats of finishing oil, rubbing between coats with pumice and lemon oil.

   *Note:* It is easier to disassemble the table for sanding and finishing.

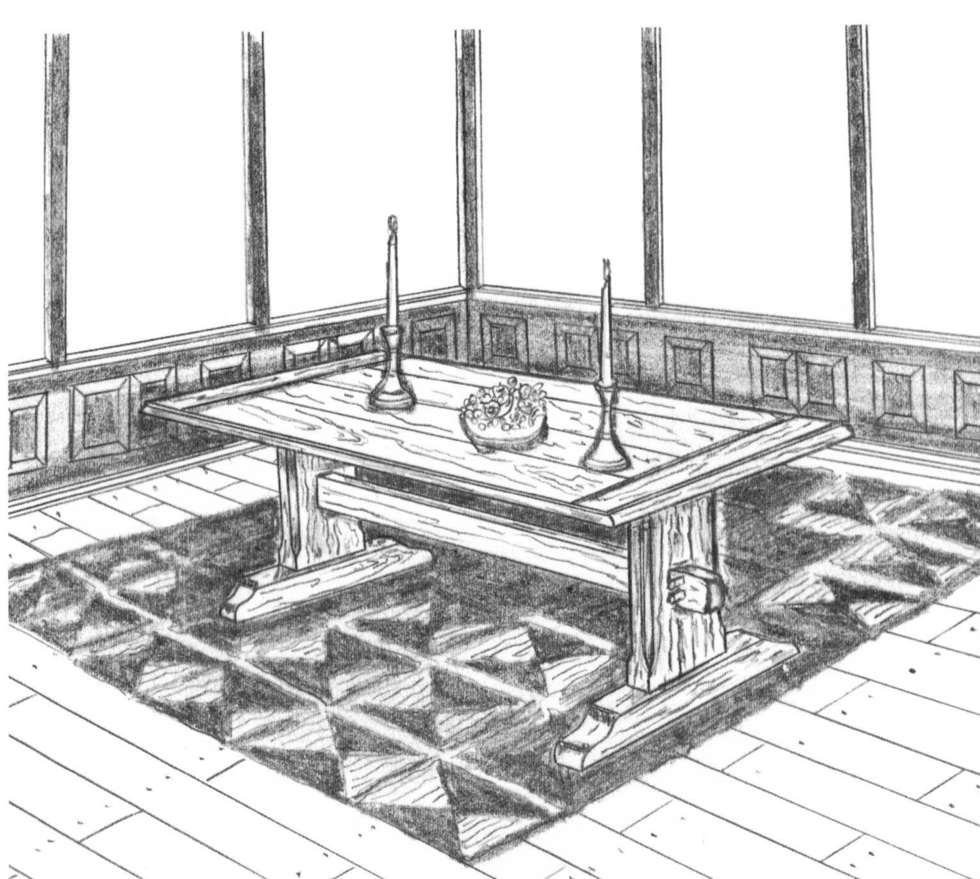

## Project 66

# STORAGE BENCH

### Materials Required

*Wood of Your Choice*

**2 sides** ¾″ × 18″ × 30″
**1 bottom** ¾″ × 16¾″ × 38½″
**1 back** ¾″ × 30″ × 38½″
**1 front** ¾″ × 14″ × 38½″
**1 section divider** ¾″ × 14″ × 15½″
**2 cleats on which lid is to rest** ¾″ × 2½″ × 15½″
**1 back board to which lid is hinged** ¾″ × 3″ × 38½″
**1 lid** ¾″ × 14½″ × 38⅜″
**1 pair H-surface hinges**
**wood screws** #9, 1½″ flathead
**wood screws** #9, ¼″ flathead

### Procedures

1. Lay out and cut all stock to size. Use bar clamps to edge-glue the stock to the necessary widths. Cut the design out of the side pieces.

   *Note:* If desired, a cyma curve can be added to the back.
2. Scrape and sand all pieces in preparation for assembly.
3. Lay out counterbored pilot holes, then use #9, 1½-inch flathead screws to fasten the back, bottom, and front to the two side pieces.

   *Note:* The bench design allows the bottom front edge to extend out approximately ¾ inch. It is attractive if the edge is shaped to a quarter-round cut. Use a router or hand plane and file for this operation.

# Storage Bench

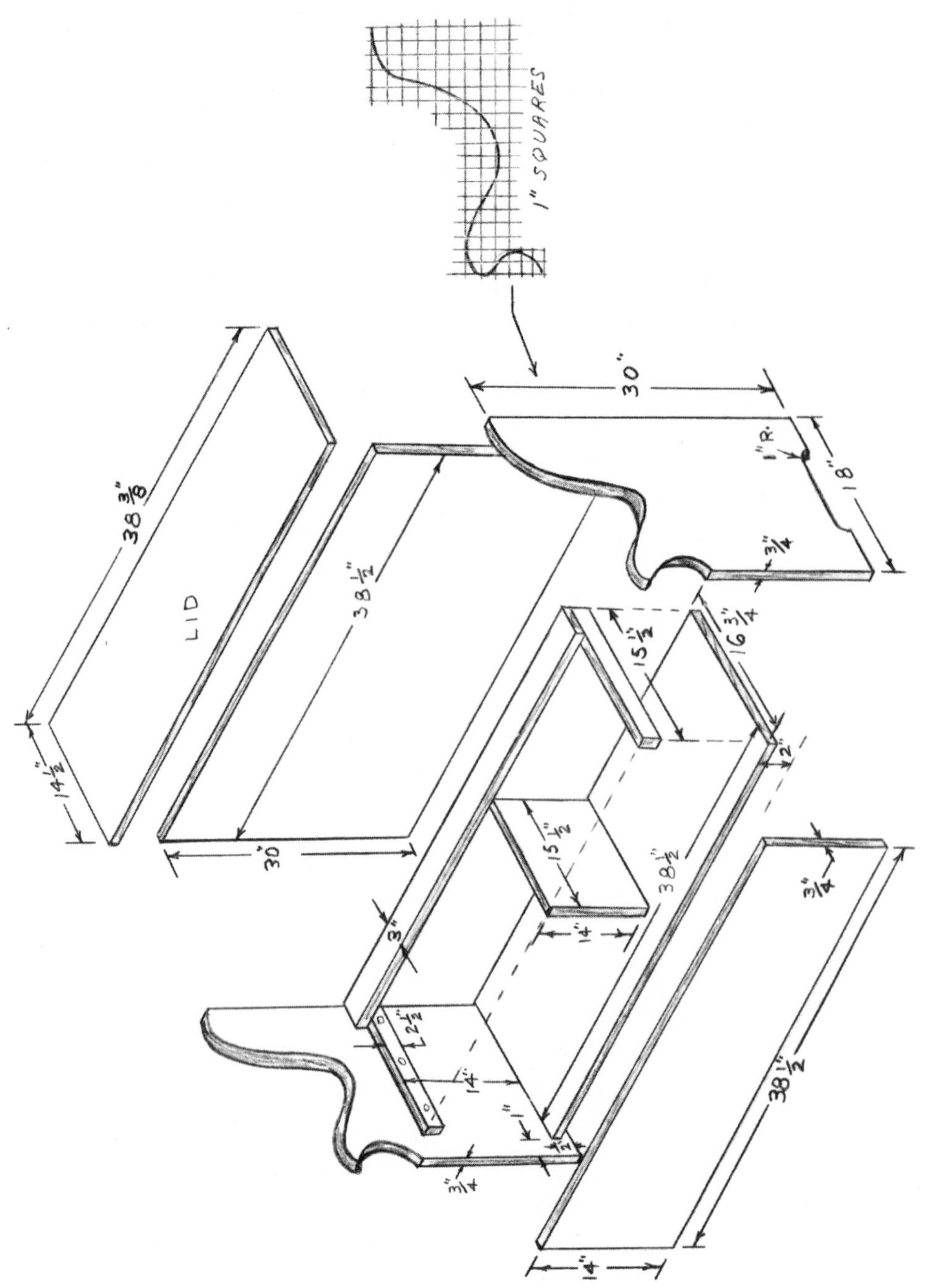

4. Attach the section divider to the exact center with glue and #6 finishing nails. Continue to drill counterbored pilot holes into the lid cleats, then fasten them to the sides with #9, 1¼-inch flathead screws.
5. Drill counterbored pilot holes through the 3-inch by 38½-inch lid hinge board, then fasten it to the rear of the cleats. Use #9, 1½-inch flathead screws.
6. Round over the front edge of the lid with a router or hand plane and file. Locate and fasten the surface H-hinges approximately 3 inches from each side piece.
7. Set and fill all holes. Use wood plugs to cover all screw heads, then sand the storage bench to its final smooth surface. Select and apply the stain or paint of your choice, then cover with three coats of polyurethane finish. Rub between coats with 320 emery paper or fine steel wool. Preserve the finish with an occasional coat of lemon oil.

STORAGE BENCH
EXCELLENT FOR STORING FIRE WOOD OR TOYS.

## Project 67

# BLACKBOARD/PHONE SHELF

### Materials Required

*Wood of Your Choice*

**1 blackboard back** ¼″ × 15½″ × 26″
**2 sides** ⅝″ × 10½″ × 27″
**1 bottom shelf** ⅝″ × 10½″ × 15½″
**1 middle shelf** ⅝″ × 10½″ × 15½″
**1 top shelf** ⅝″ × 2½″ × 15½″
**1 top decorative piece** ⅝″ × 2½″ × 15″
**1 bottom back piece** ⅝″ × 1⅝″ × 15″
**1 key rack** ⅝″ × 1⅝″ × 15″
**brads** ¾″ finishing
**6 cup hooks** ¾″ dia.
**blackboard paint** Available at your local hardware or paint supply store.

### Procedures

1. Lay out and cut all pieces to the suggested size and design.
2. Lay out and cut the three ¼-inch by ⅝-inch dado joints.
3. Lay out and cut the ¼-inch by ⅜-inch rabbet joint on the back edge of both side pieces. The rabbet cut is made to recess the blackboard.
4. Fasten the three shelves to the dado joints in the side pieces by gluing and clamping.
5. Make the 15½-inch by 26-inch blackboard by cutting a piece of hardboard then applying two coats of special blackboard paint to the smooth side of the hardboard.
6. Fasten the blackboard to the back by gluing and nailing into the rabbet joints.

# Blackboard/Phone Shelf

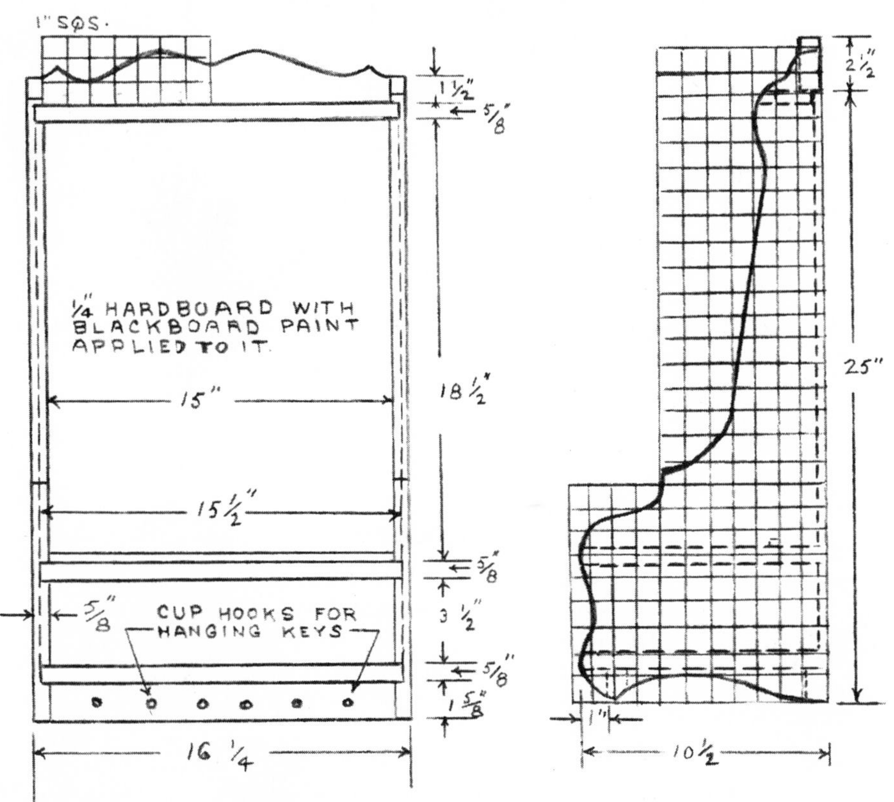


7. Attach the top decorative piece to the top shelf by gluing and clamping. Locate and fasten the cup hooks to the 1⅝-inch stock, then glue the key rack to the bottom shelf.
8. Scrape off all traces of glue. Set and fill all nail holes, then sand the project to a smooth surface.
9. Select and apply the colored stain or paint of your choice. When dry, apply several coats of lacquer. Rub between coats with fine steel wool or pumice and oil. Protect the finish by applying a coat of paste wax and buffing.

## Project 68

# RECORD CABINET

### Materials Required

*Wood of Your Choice*

**2 sides** ¾″ × 14″ × 32″
**1 top shelf** ¾″ × 13″ × 25¼″
**1 middle shelf** ¾″ × 13¾″ × 25¼″
**1 bottom shelf** ¾″ × 14″ × 25¼″
**1 top design piece** ¾″ × 2½″ × 24½″
**1 front skirt** ⅝″ × 4″ × 26″
**1 back** ¼″ × 25¼″ × 25¼″
**6 dowels** ⅜″ dia. × 25¼″

### Procedures

1. Use bar clamps to edge-glue the sides and shelves.
2. Cut square the three shelves and two sides as shown in the drawing.
3. On the rear inner edge of the top and bottom shelf along with the two sides cut a ¼-inch by ⅜-inch rabbet to receive the back piece.
4. Lay out and cut the three ⅜-inch by ¾-inch dadoes into each side board to receive the three shelves.
5. Accurately lay out the ⅜-inch-diameter holes for the dowels to divide the records. All holes should be exactly on top of each other at 6⅛-inch intervals on all three shelves.

   Drill the eighteen ⅜-inch-diameter holes. The ⅜-inch holes in the lower two shelves are to be drilled completely through while the holes in the underside of the top shelf are to be to a depth of ⅜ inches.

# Record Cabinet

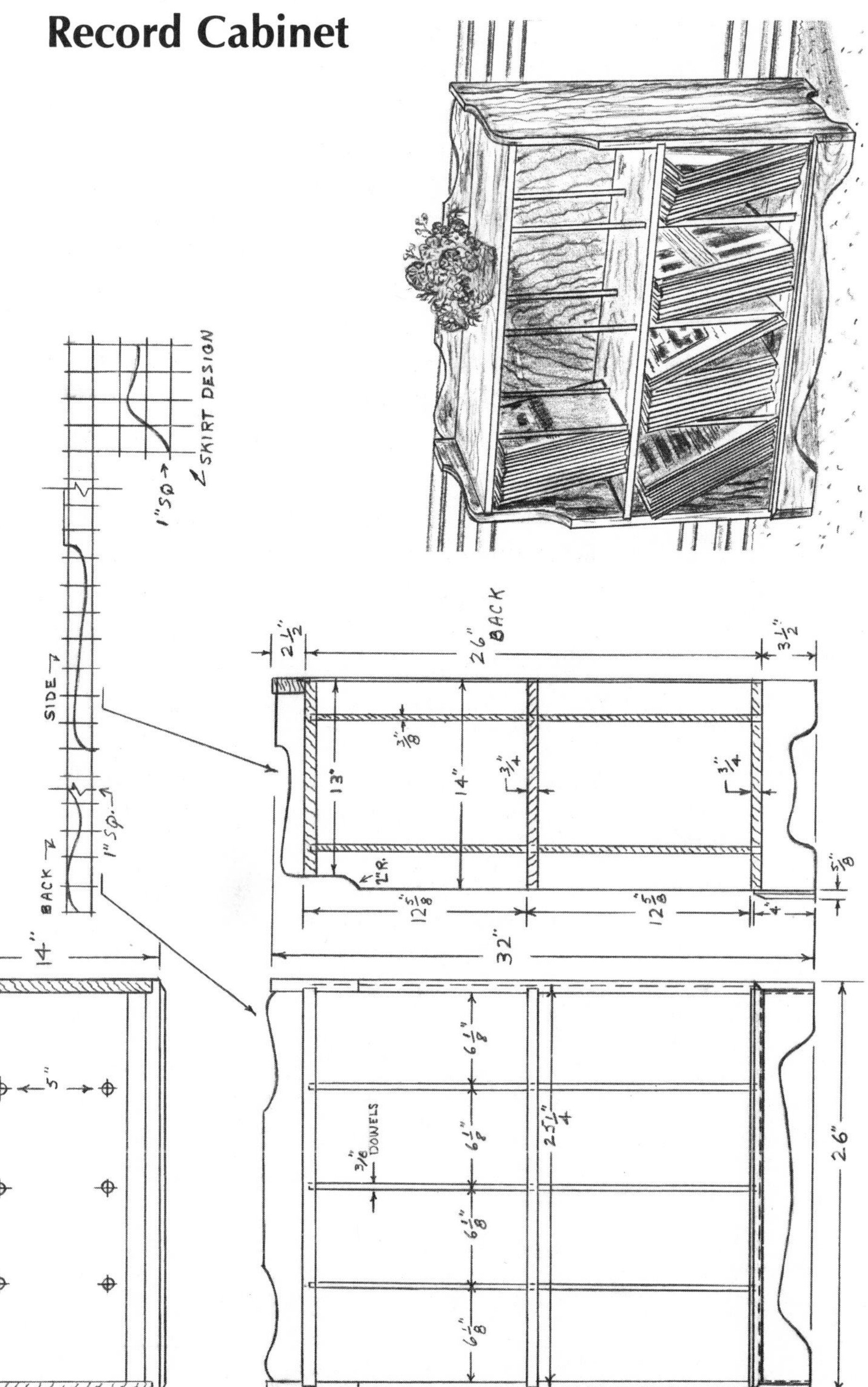

6. In preparation for the case assembly scrape and sand all of the pieces smooth.
7. Fasten the sides into the dadoes and the back into the precut rabbets with glue and #6 finishing nails. Before permanently attaching the back with ⅞-inch flathead brads, check the case assembly for squareness (90 degrees) with a carpenter's square.

   Insert the six ⅜-inch-diameter by 25¼-inch dowels up through the bottom shelf and into the remaining predrilled holes.

   Insert glue into the holes drilled on the underside of the top and secure the dowels in place.
8. Lay out and cut the designs on the top back piece along with the skirt. Attach the two pieces into position by gluing and clamping.
9. With a file and sandpaper round over all exposed edges approximately ⅛ inch.
10. Set and fill all holes, scrape off all traces of glue, then sand the project to a final smooth surface. Select and apply the stain or paint of your choice, then cover with several coats of polyurethane finish. Rub between coats with 320 emery paper. Preserve the finish with an occasional coat of lemon oil.

Project 69

# ADJUSTABLE BOOKCASE

## Materials Required

*Wood of Your Choice*

**1 top** ¾″ × 13″ × 38″
**1 bottom** ¾″ × 12¾″ × 35″
**2 adjustable shelves** ¾″ × 12″ × 34⅛″
**2 sides** ¾″ × 12¾″ × 35¼″
**1 front skirt** ¾″ × 3¼″ × 34¼″
**1 top rail** ¾″ × 1″ × 34¼″
**2 cleats** ¾″ × ¾″ × 10½″
**1 back** ¼″ × 35″ × 31½″
**8 dowel pegs for support of shelf** ½″ dia. × 1¼″

## Procedures

1. Using bar clamps, edge-glue enough stock to meet the required width for all pieces necessary.
2. Lay out and cut square (90 degrees) the top, bottom, and two sides to the size and design indicated.
3. On the rear inner edge of the top, bottom, and two sides cut a ¼-inch by ⅜-inch rabbet to receive the back.
4. Cut ⅜-inch by ¾-inch dadoes 3¼-inches up from the bottom of the side pieces to receive the bottom shelf.
5. Drill countersunk pilot holes, then fasten the ¾-inch by 10½-inch cleats even to the top edge of the two side pieces. Use #10, 1¼-inch flathead screws.

# Adjustable Bookcase

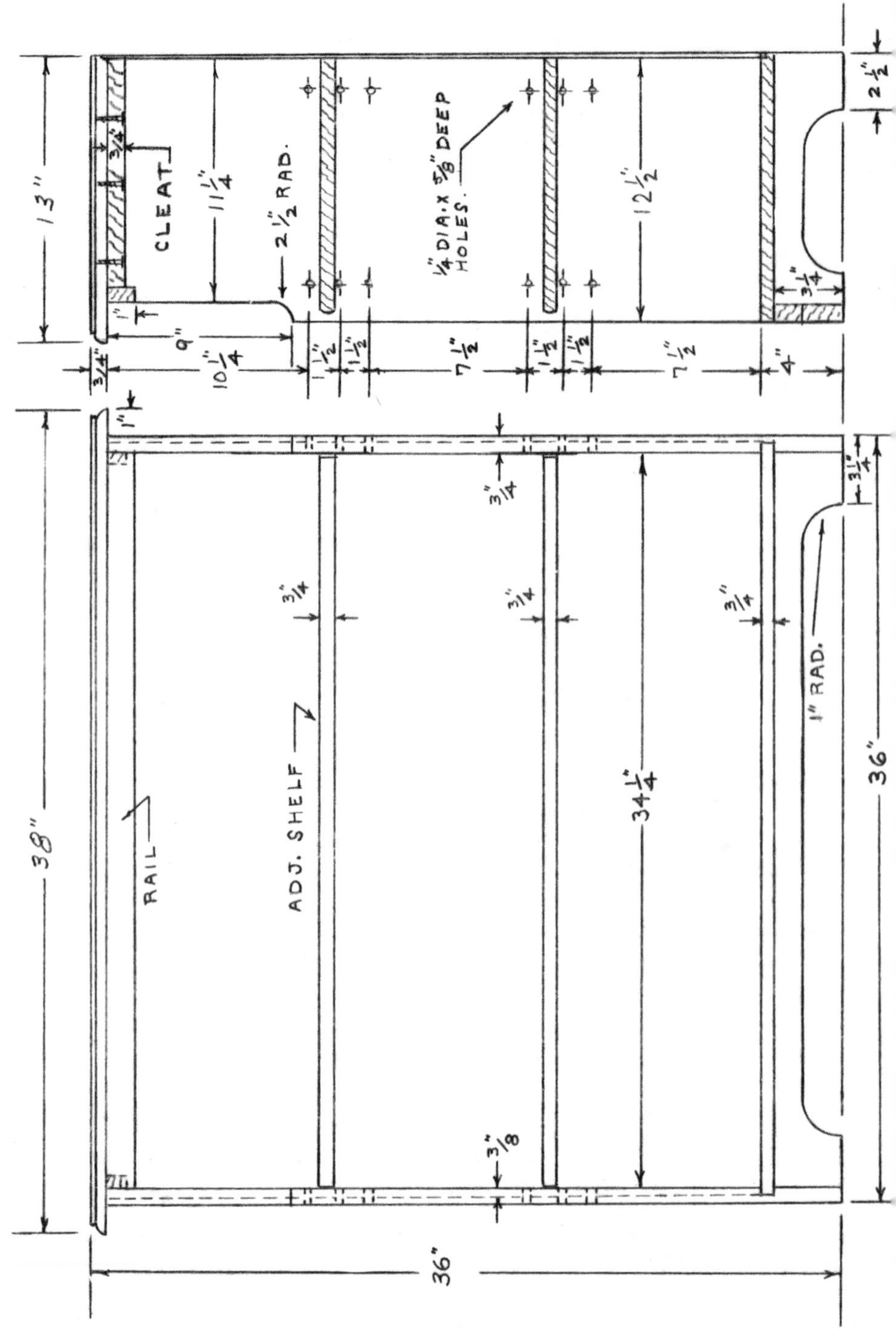

6. Accurately lay out the ¼-inch-diameter by ⅝-inch-deep holes for the adjustable shelves, then drill these holes absolutely straight. Use a try square to guide the drill to a 90-degree angle.

7. Before starting the case assembly, scrape and sand the pieces smooth.

   Attach the top into place by drilling countersunk pilot holes up through the cleats and into the bottom side of the top. Use #10, 1¼-inch screws to secure the top to the cleats.

   Fasten the bottom shelf into the precut dadoes with glue and #6 finishing nails. Proceed to install the back into the precut rabbets using ⅞-inch flathead brads.

   *Note:* Before the back is nailed permanently, check the case assembly for squareness (90 degrees).

8. Secure the 1-inch by 34¼-inch rail to the front edge of the cleats and to the underface of the top by using glue and #6 finishing nails.

9. Cut the skirt to the correct design and size, then attach the skirt to the sides and underface of the bottom shelf using glue and #6 finishing nails.

10. Lay out and cut the two adjustable shelves to size. With a router or plane and file, round over the front edge of the shelves.

11. Use a router and quarter-round bit or roman ogee bit to shape the three exposed edges of the top.

12. Cut eight ¼-inch-diameter by 1¼-inch birch dowels for the shelf supports. Round each end slightly.

13. Set and fill all holes, then sand the bookcase to a smooth surface. Select and apply the paint or stain of your choice, then cover with three coats of clear lacquer. Rub the final coat down with pumice and lemon oil. Preserve the finish with an occasional coat of lemon oil.

## Project 70

# BUFFET

### Materials Required

*Wood of Your Choice*

**1 top** ¾″ × 20″ × 55″
**2 sides** ¾″ × 18½″ × 36¼″
**4 solid drawer divider dust panels with shelf and bottom** ¾″ × 18¼″ × 51¼″
**1 top drawer middle divider** ¾″ × 4½″ × 18″
**2 front cabinet stiles** ¾″ × 4″ × 19½″
**1 front divider stile** ¾″ × 6″ × 19½″
**1 skirt, front** ¾″ × 5¼″ × 53½″
**2 skirts, sides** ¾″ × 5¼″ × 19¼″
**1 back** ¼″ × 36⅝″ × 51¼″
**1 front top rail** ¾″ × ¾″ × 50½″
**2 top cleats** ¾″ × ¾″ × 16″

*One Door:*

**2 stiles** ¾″ × 3″ × 20¼″
**1 upper crown rail** ¾″ × 8″ × 13″
**1 lower rail** ¾ × 3″ × 13″
**10 dowel pins** ⅜″ dia. × 2″
**1 panel** ¼″ × 15″ × 16″

**1 top drawer unit, including front overlap (overall size)** 4″ × 26″ × 17¼″
**1 bottom drawer unit, including front overlap (overall size)** 5″ × 50½″ × 17¼″

### Procedures

1. Use bar clamps to edge-glue enough stock to meet the required widths for all pieces.

# Buffet

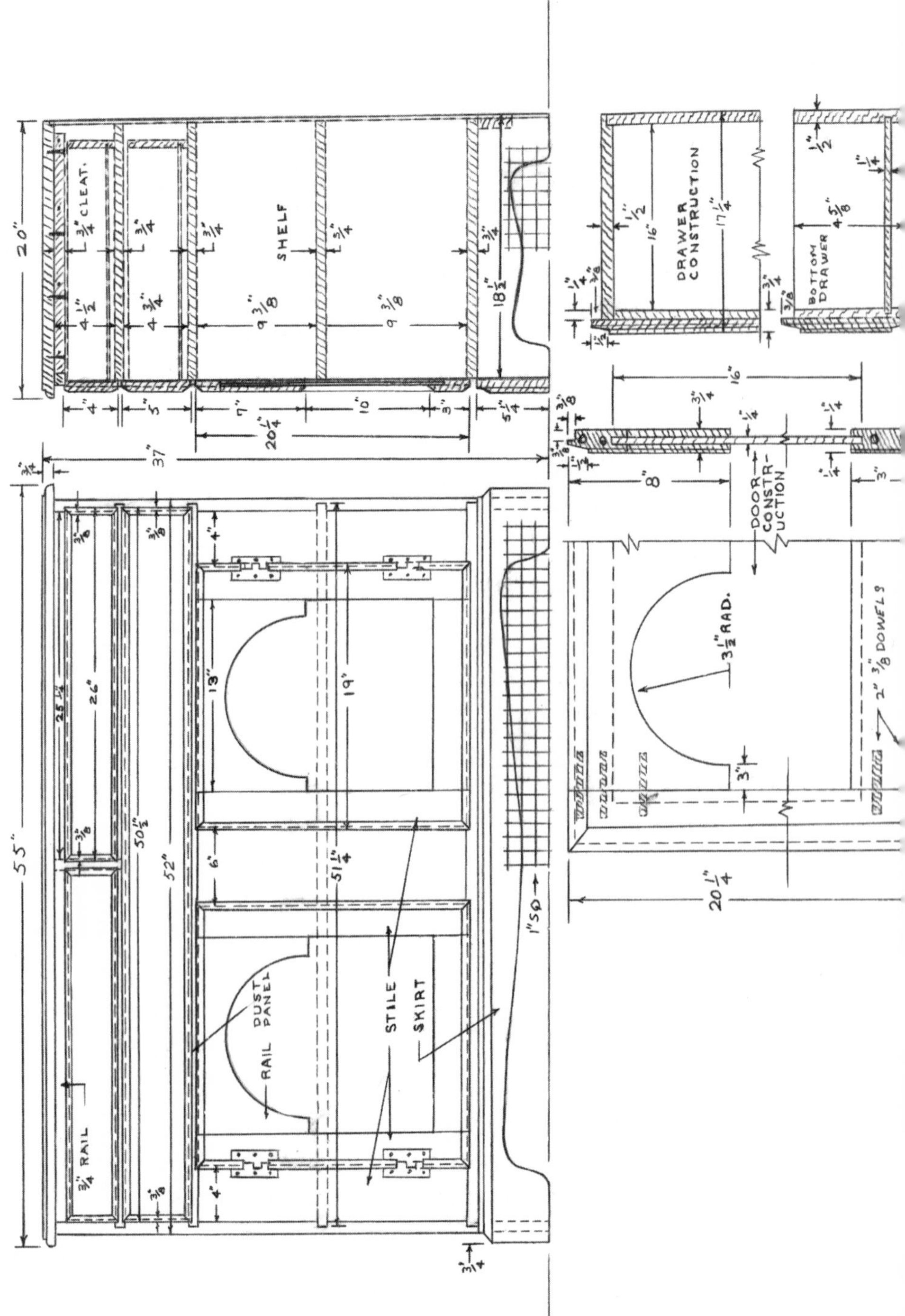

3/4" CLEAT
SHELF
DRAWER CONSTRUCTION
BOTTOM DRAWER
DOOR CONSTRUCTION
3 1/2" RAD.
2" 3/8 DOWELS
1" SQ
DUST PANEL
RAIL
STILE
SKIRT
3/4 RAIL

2. Lay out and cut the two sides to the size indicated in the drawing.

3. Cut the ⅜-inch by ¾-inch dadoes on the inner faces of the two sides to receive the drawer dividing rails, shelf, and bottom.

   Cut ¼-inch by ⅜-inch rabbets on the back inside edge of the side pieces to receive the back.

4. Cut the two solid dust panel drawer dividers, the shelf, and bottom to the size indicated.

   Lay out and cut the top to the correct size.

   Shape the front and two side edges of the top with a router and quarter-round bit; leave a ⅛-inch lip.

   Cut a ¼-inch by ⅜-inch rabbet on the back inner edge of the top to receive the back.

5. Drill countersunk pilot holes, then fasten the ¾-inch by ¾-inch by 16-inch cleats flush to the top edge of the side pieces. Use #10, 1¼-inch wood screws.

6. Before assembling, smooth the pieces by planing, scraping, and sanding.

   Fasten the cabinet frame together by gluing and clamping with bar clamps. The top is held in place by 1¼-inch screws installed through the cleats and into the inner face of the top.

   *Note:* Check the structure for squareness before the glue sets up.

7. Cut square (90 degrees) the ¼-inch back, then proceed to fasten it to the precut rabbets. Use 1-inch flathead brads.

8. Install the top ¾-inch by ¾-inch by 50½-inch rail to the front edge of the cleats and to the underface of the top piece. Use glue and wooden screw clamps.

Cut a ¾-inch by ¾-inch blind rabbet into the front top edge of the middle drawer divider. The cutout is made so that the middle drawer divider will fit flush to the front edge of the top dust panel. Install with glue and #6 nails. Make sure it is exactly on center.

9. Construct each drawer unit to the dimensions in the drawing. Use the construction hints found in the front of the book. Be sure the front of each drawer unit has a ⅜-inch overlap on the inner edge along with a chamfer or beveled cut on the outer edge.

10. Lay out and cut all of the pieces necessary to make the two paneled doors.

    Before cutting the design in the top door rails, cut a ¼-inch by 4-inch groove or rabbet into the top rails along with a ¼-inch by 1-inch groove or rabbet into the remaining rails and stiles.

    Lay out and cut the radius in the top rail to form the crown design. Make dowel or mortise and tenon joints to fasten the rails to the stiles. Before assembling the doors, use a router and chamfer bit to shape the outside edges.

    Insert the ¼-inch pieces into the precut grooves or rabbet, then fasten the door corners together with ⅜-inch by 2-inch dowels to form the doweled joints.

    Cut a ⅜-inch by ⅜-inch overlap rabbet completely around the inner edge of the doors so that they can fit over the door openings.

    Attach the doors to the stiles with ⅜-inch offset surface H-hinges. Install 1¼-inch-diameter brass knobs on the doors.

11. Lay out and cut the three skirt pieces to the indicated size and design.

    Cut miter joints for the corners, then fasten the skirt in place with glue and #6 finishing nails.

    *Note:* Cut a section out of the base of each side piece so that when the skirt is attached, the inner side will not show through the design in the skirt.

12. Set and fill all holes, then sand the buffet to a smooth surface. Select and apply the stain of your choice, then cover with a minimum of three coats of lacquer. Preserve the finish with an occasional coat of paste wax or lemon oil.

## Project 71

# CHEST

### Materials Required

*Wood of Your Choice*

**1 top** ¾″ × 13″ × 22″
**1 base** ¾″ × 13″ × 22″
**2 sides** ¾″ × 7″ × 20¾″
**2 butt hinges** ⅛″ × ¾″
**nails** #6 finishing

### Procedures

1. Glue up and clamp the stock to the required width, then lay out and cut the pieces to the suggested size.
2. With a router and quarter-round bit cut the design around the top and base pieces.
3. Lay out and cut the ⅜-inch by ¾-inch rabbet joints on the front and back pieces.
4. Fasten the sides together by gluing and clamping. Fasten the base to the box assembly with glue and #6 finishing nails.
5. Locate and fasten the two ¾-inch butt hinges approximately 2 inches from each end.
6. Scrape off all traces of glue. Set and fill all nail holes, then sand the chest to a smooth surface.
7. Select and apply the stain or paint of your choice. When dry, apply several coats of clear lacquer finish, rubbing between coats with fine steel wool or pumice and lemon oil. To protect the finish apply a coat of paste wax and buff.

# Chest

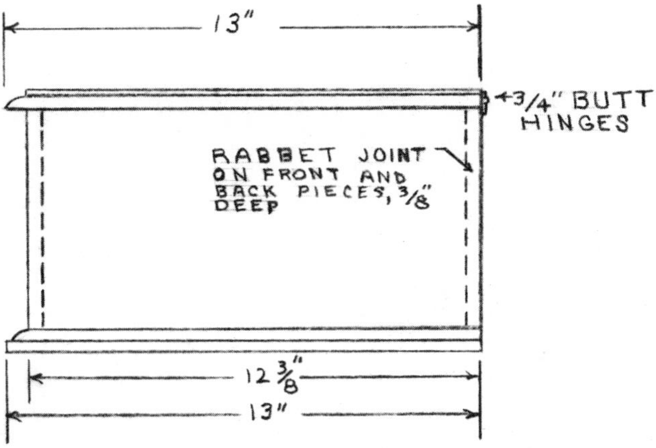

13"

3/4" BUTT HINGES

RABBET JOINT ON FRONT AND BACK PIECES, 3/8" DEEP

12 3/8"

13"

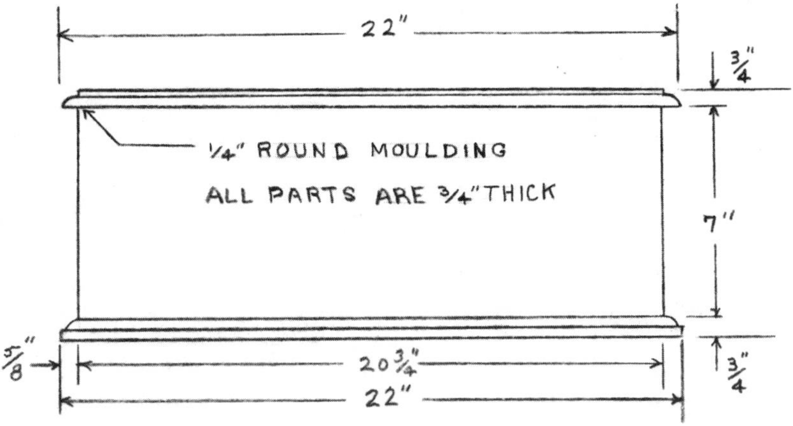

22"

3/4

1/4" ROUND MOULDING

ALL PARTS ARE 3/4" THICK

7"

5/8"

20 3/4"

3/4

22"

## Project 72

# HUTCH

### Materials Required

*Wood of Your Choice*

*Top Section:*
**1 top** ¾″ × 11½″ × 38″
**1 cove corner molding** ¾″ × 1″ × 62″
**2 sides** ¾″ × 10″ × 37¾″
**2 shelves** ¾″ × 9″ × 34¼″
**1 back** ¼″ × 34¼″ × 38″
**1 door header** ¾″ × 2″ × 33½″
**4 dowels** ⅜″ dia. × 2½″

*Top Section Door Frames:*
**4 stiles** ¾″ × 2½″ × 25″
**4 rails** ¾″ × 2½″ × 16⅝″
**2 glass sheets** ⅛″ × 19⅞″ × 19¼″
**4 antique surface H-hinges** 2½″
**knobs of your choice**

*Bottom Section:*
**1 top** ¾″ × 17½″ × 37½″
**2 sides** ¾″ × 16½″ × 33¾″
**1 back** ¼″ × 34¼″ × 31″
**1 shelf** ¾″ × 15½″ × 34¼″
**1 bottom** ¾″ × 15½″ × 34¼″
**1 base molding to front** ¾″ × 3¼″ × 36½″
**2 base moldings to sides** ¾″ × 3¼″ × 17¼″
**1 door header** ¾″ × 1″ × 33½″
**2 doors** ¾″ × 16⅝″ × 30½″
**2 cleats** ¾″ × 1″ × 14¾″
**4 antique surface H-hinges** 4″
**knobs of your choice**

# Hutch

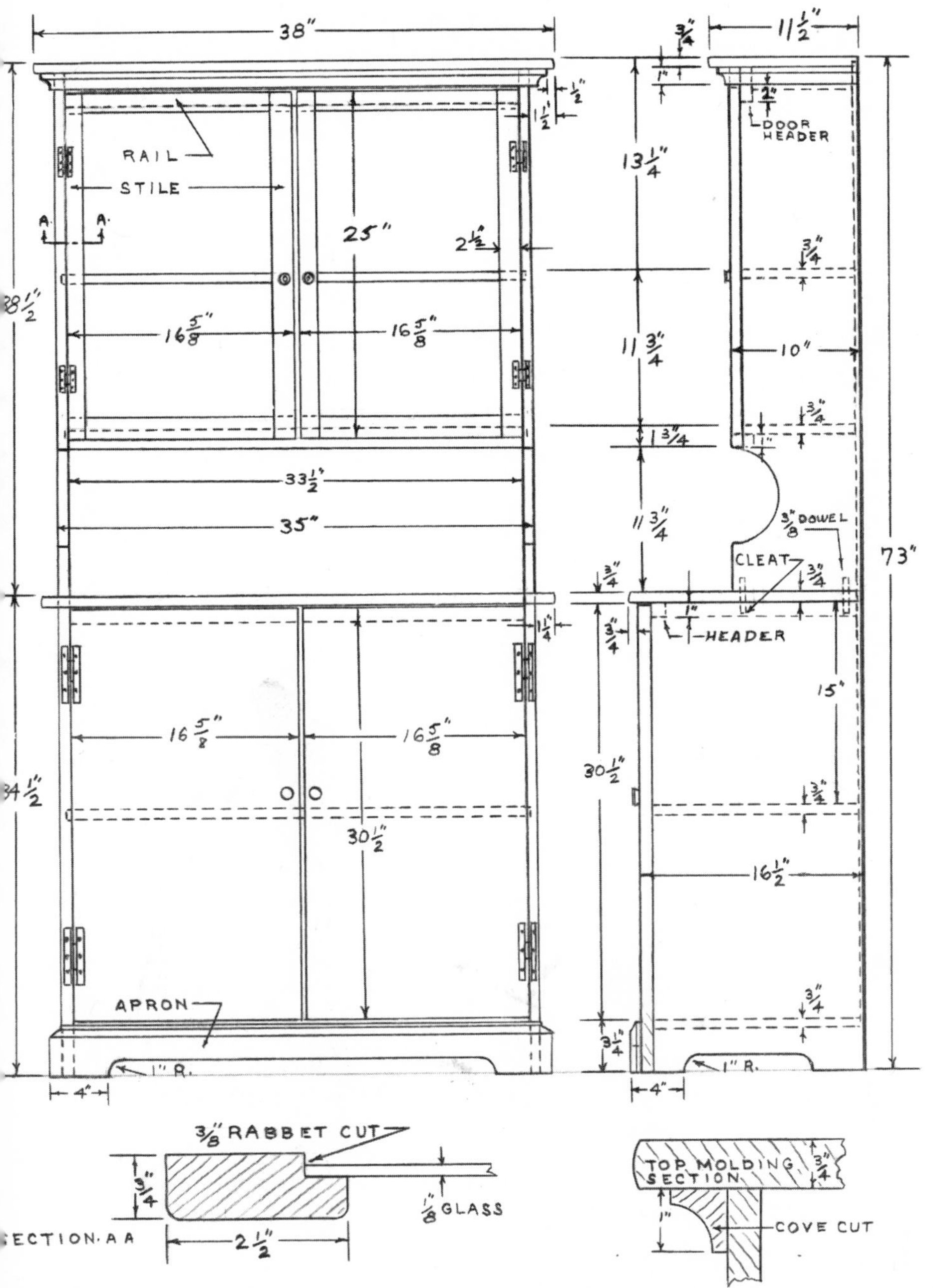
38"
11 1/2"
RAIL
STILE
A. A.
25"
2 1/2"
16 5/8"
16 5/8"
13 1/4"
11 3/4"
1 3/4
11 3/4
DOOR HEADER
10"
3/8" DOWEL
CLEAT
33 1/2"
35"
HEADER
15"
16 5/8"
16 5/8"
30 1/2"
30 1/2"
16 1/2"
73"
APRON
1" R.
1" R.
4"
4"
3/8" RABBET CUT
1/8" GLASS
SECTION A A
2 1/2"
TOP MOLDING SECTION
COVE CUT

## Procedures

1. This cabinet is made in two separate sections. Begin by laying out, cutting to length, and edge-gluing enough stock to complete all of the pieces necessary to construct the cupboard. Be sure all of the stock is cut square (90 degrees).

2. Cut ⅜-inch by ¾-inch dadoes on the inner faces of the four side pieces to receive the two lower shelves and the two upper shelves.

   Cut ¼-inch by ⅜-inch rabbets on the rear inside edges of each of the four side pieces and the two top pieces. This cut is to receive the ¼-inch back.

3. Smooth the pieces by hand-scraping, planing, and sanding. Proceed to glue and nail the four shelves into the precut dadoes in the top and bottom sections.

4. Drill countersunk pilot holes, then fasten ¾-inch by 1-inch cleats flush with the top edge of each of the four side pieces with #10, 1¼-inch flathead wood screws.

5. Round over the edges of the two top pieces along the three exposed sides with a portable router or a hand plane and file.

   Attach the tops above each cabinet section by drilling counterbored pilot holes from the underside of the cleats and into the underside of top pieces. Use #10, 1½-inch flathead wood screws.

6. Glue and clamp the bottom section door header approximately 1½ inches back from the top front edge.

   Fasten the header to the top section of the cabinet approximately 1½ inches back from the front edge.

Fasten a ¾-inch by 1-inch by 33½-inch rail below the bottom shelf in the top section approximately ¾ inch back from the front edge of the sides. Use glue and clamps. Reinforce with #4 finishing nails.

7. Cut out the design for the base skirt with a sabre saw, then make a ½-inch chamfer or cove cut along the top edge.

   Cut 45-degree miter joints for the two corners, then proceed to fasten the skirt to the cabinet base by using glue and #6 finishing nails. The chamfer can be cut with a portable router or hand plane.

8. Cut a ¾-inch cove design on the edge of a ¾-inch by 1-inch piece of stock to make the molding to fit beneath the top piece of the top section.

   Cut 45-degree miters for the corners, then attach in place with glue and #6 finishing nails.

   *Note:* Cove molding can be bought premade at your local hardware store.

9. Lay out and cut square the two ¼-inch backs, then fasten them to the precut rabbets on the rear edges with 1-inch brads.

   *Note:* Make sure both cabinet frames are square (90 degrees) before fastening the backs permanently.

10. Cut all parts to the upper section doors. Make end half-lap joints or mortise and tenons to fasten the corners.

    Cut a ⅜-inch by ⅜-inch rabbet on the inside edges of each part to receive the ⅛-inch glass.

Assemble the door frames by gluing and clamping. Round the outside edges with a router and ¼-inch round bit.

11. Fit the doors for the upper and lower sections into position, then fasten into place with surface hinges. Use magnetic catches placed on the headers to keep the doors closed.
12. Drill eight ⅜-inch holes completely aligned to the top and bottom sections to receive the dowels to hold the top of the cabinet onto the bottom section of the cabinet.
13. Sand the cupboard smooth, then apply the finish of your choice, rubbing between coats with 320 emery paper. Use paste wax to preserve the finish.

## Project 73

# COUNTRY DRY SINK

### Materials Required

*Wood of Your Choice*

**1 top** ¾″ × 9″ × 30″
**1 top shelf** ¾″ × 12¾″ × 28″
**1 bottom shelf** ¾″ × 11½″ × 26½″
**2 upper sides** ¾″ × 7″ × 14½″
**2 lower sides** ¾″ × 13″ × 20¼″
**1 back to upper section** ¾″ × 7″ × 29″

*Back Paneled Section:*
**1 top rail** ¾″ × 2″ × 26½″
**1 base rail** ¾″ × 4¼″ × 26½″
**1 panel** ¼″ × 16″ × 26½″
**1 front to top section** ¾″ × 4″ × 29″
**1 skirt** ¾″ × 4″ × 28″
**2 stiles** ¾″ × 5½″ × 16¼″
**1 door** ¾″ × 17″ × 16⅛″
**1 door header** ¾″ × 1½″ × 26½″
**2 cleats** ¾″ × 1½″ × 11¾″
**2 antique surface H-hinges**
**1 brass door knob**

### Procedures

1. The dry sink is made in two sections. Begin by laying out, cutting to length, and edge-gluing with bar clamps enough stock for all of the necessary pieces. Be sure all of the stock is cut square (90 degrees).
2. Lay out and cut the design on the upper side pieces, making sure the front edges of the sides are angled back approximately 75 degrees. Use a protractor and T-bevel to check the angles for accuracy.

# Country Dry Sink

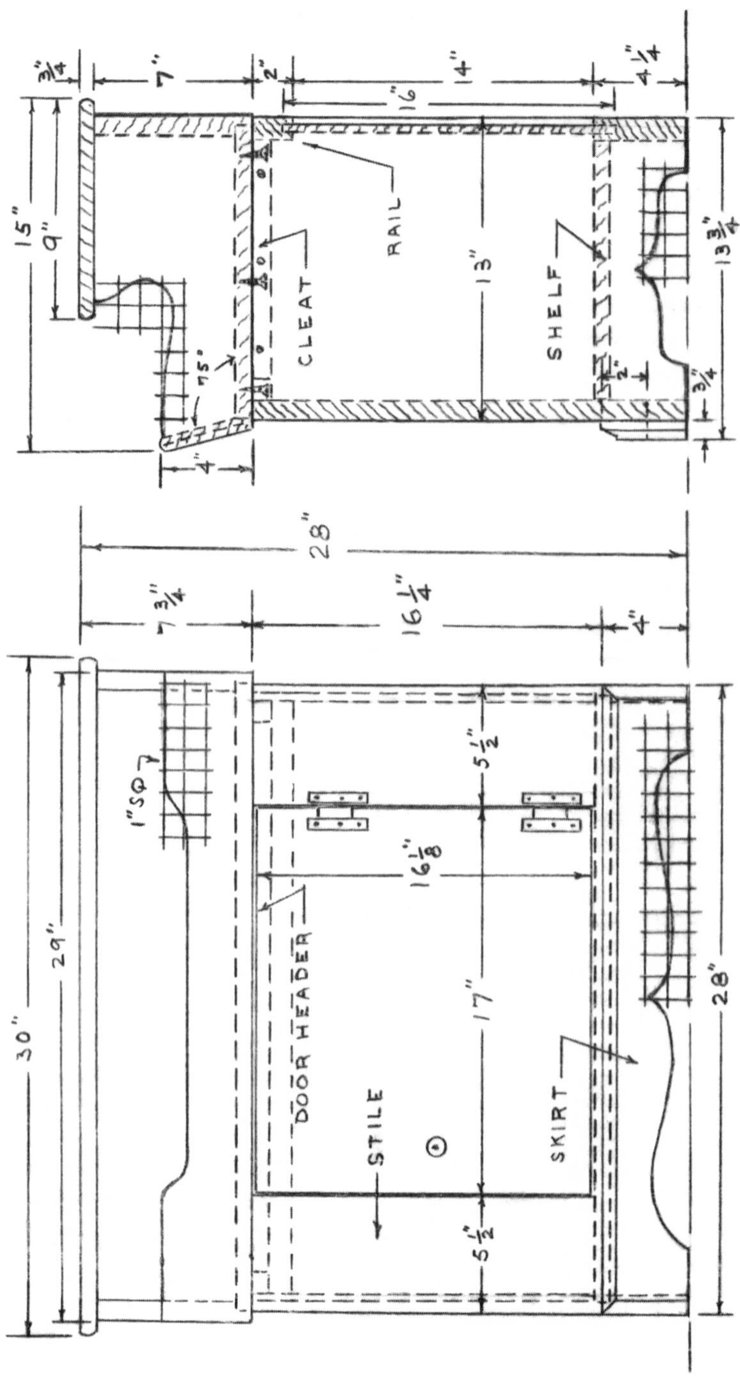

Lay out and cut two ⅜-inch by ¾-inch rabbets along the bottom edge of each side piece.

3. Make the design on the upper front piece, then cut it to shape.

   Cut two ⅜-inch by ¾-inch rabbets at each end of the front design piece.

   Lay out and cut a ¾-inch chamfer at approximately a 25-degree angle along the inner bottom of the front piece to allow for a flush fit to the shelf. Cut chamfer with a hand plane or an electric router and chamfer bit.

4. Cut the upper back piece to size, then make a ⅜-inch by ¾-inch rabbet cut along the inner bottom edge.

   Cut the 9-inch-wide top to size, then round over all edges with a plane and file or portable router and quarter-round bit.

5. Smooth all the pieces by hand-scraping, filing all edges, hand-planing, and sanding.

   Assemble the upper section's six pieces using glue and #6 finishing nails.

   *Note:* Before the glue hardens, check for squareness (90 degrees).

6. Lay out and cut the design on the two sides for the bottom section, then proceed to cut the ⅜-inch by ¾-inch blind dadoes on the two sides to receive the lower shelf.

7. Construct the back panel assembly by cutting a ¼-inch by ½-inch groove into the edge of the top 3-inch by 26½-inch rail and into the edge of the 4¼ by 26½-inch base rail. Insert the ¼-inch by 16-inch by 26½-inch panel into the grooves with glue.

8. From a pattern, cut out the design to the front skirt. Make a ½-inch chamfer cut on the three exposed edges. Use a hand plane or an electric router to cut the chamfer.

9. Cut the ¾-inch by 1½-inch by 26½-inch door header to size, then cut the two ¾-inch by 1½-inch by 11¾-inch cleats.

   Fasten the cleats flush to to the top edge of the side pieces. Cleats should be placed 1½ inches from the front edge and ¾ inch from the rear edge of the sides. Use countersunk pilot holes and #10, 1¼-inch flathead screws to fasten.

   *Note:* The header will be fastened to the front of the cleats with #10, 1½-inch flathead wood screws counterbored and plugged.

10. Lay out and cut the two ¾-inch by 5½-inch by 16¼-inch stiles to size. The stiles should be cut in at an angle at the bottom to match the skirt's cyma curve design.

11. After carefully filing, sanding, and scraping smooth all pieces, begin the bottom section assembly by attaching the shelf into the precut dadoes in the side pieces, then fasten the back paneled section to the rear inner face of the side.

    Fasten the stiles to the front of the frame, then secure a ¾-inch by 2-inch by 17-inch piece of stock between the two stiles and flush to the top edge of the skirt. This piece is used to fill the gap left between the shelf and the skirt.

    Mount the door to the right stile with antique surface hinges.

    Finish the bottom assembly by gluing and nailing the skirt in place. Use glue and #6 finishing nails as fasteners.

12. Attach the section of the cabinet to the bottom section by drilling countersunk pilot holes up through the cleats and into the bottom face of the top shelf. Use #10, 1½-inch flathead screws.
13. Give a final sanding, then apply the finish of your selection. Rub between coats of clear finish with #320 emery paper or fine steel wool. Preserve the finish with a coat of paste wax.

## Project 74

# KITCHEN UTENSIL RACK

### Materials Required

*Wood of Your Choice*

**1 back** ½″ × 13¼″ × 14¾″
**2 strips** ⅜″ × ¾″ × 13¼″
**2 dividing blocks** ¼″ × ¾″ × ⅞″
**Wood screws** #5, ¾″ flathead

### Procedures

1. Lay out and cut all pieces to the suggested size and design.
2. Glue and clamp the two ⅜-inch by 13¼-inch strips to the two ¼-inch by ⅞-inch pieces.
3. Attach the utensil holder to the back piece by gluing and clamping. When the glue dries, reinforce the holder by drilling two countersunk pilot holes through the back and into the holder. Screw in two #5, ¾-inch flathead wood screws.
4. To hang the rack drill two ⅜-inch holes to a depth of ⅝ inch. The holes should be aligned and 11 inches apart through the back of the rack.
5. Clean all traces of glue, then sand the utensil rack to a smooth finish.
6. Select and apply the stain or paint of your choice. When dry, apply several coats of clear finishing oil. Rub between coats with fine steel wool or pumice and oil. Protect the surface with a coat of paste wax and buff.

# Kitchen Utensil Rack

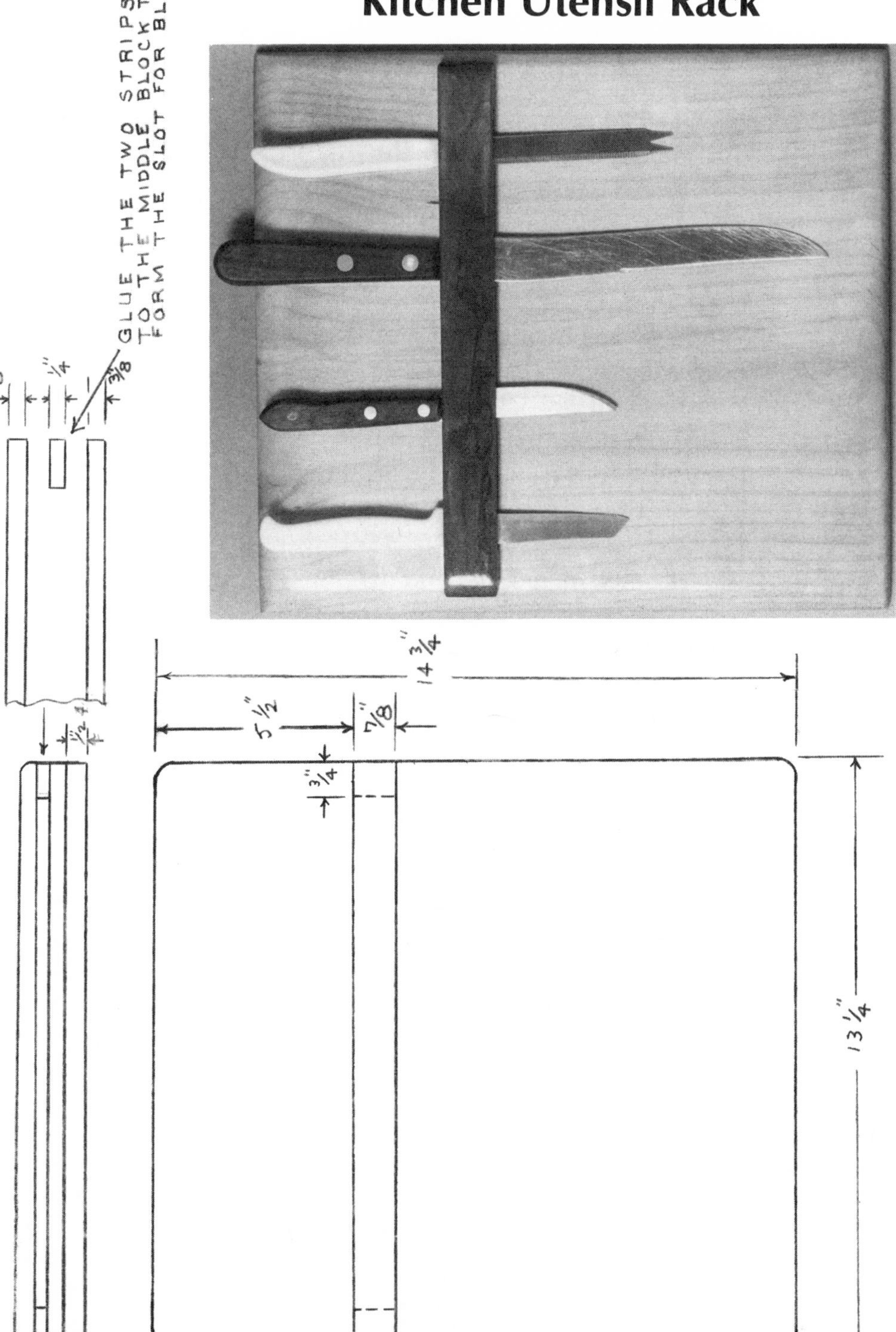

## Project 75

# CHEST OF DRAWERS

### Materials Required

*Wood of Your Choice*

**2 sides** ¾″ × 16″ × 32″
**1 top** 1″ × 18″ × 44″
**3 solid drawer divider dust panels** ¾″ × 15¾″ × 41¼″
**1 top rail** ¾″ × 1½″ × 41¼″
**1 bottom rail** ¾″ × 1½″ × 40½″
**1 skirt, front** ¾″ × 5½″ × 43½″
**2 skirts, sides** ¾″ × 5½″ × 16¾″
**1 top drawer divider** ¾″ × 8″ × 15¾″
**4 drawer guides** ½″ × 1″ × 15¾″
**drawer units, including front overlap (overall size)** 6⅜″ × 19¾″ × 15″
**drawer units, including front overlap (overall size)** 9″ × 41¼″ × 15″

### Procedures

1. Using bar clamps edge-glue enough stock to meet the required width for all pieces.

2. Lay out and cut the top and two sides to the shape and size indicated in the drawing. On these three pieces cut a ¼-inch by ⅜-inch rabbet on the rear inside edges to receive the back.

   Lay out and cut the three ⅜-inch by ¾-inch dadoes into each side piece to receive the dividing dust panels.

   At the top front edge of each side piece cut a ¾-inch by 1½-inch stop rabbet to receive the top rail.

# Chest of Drawers

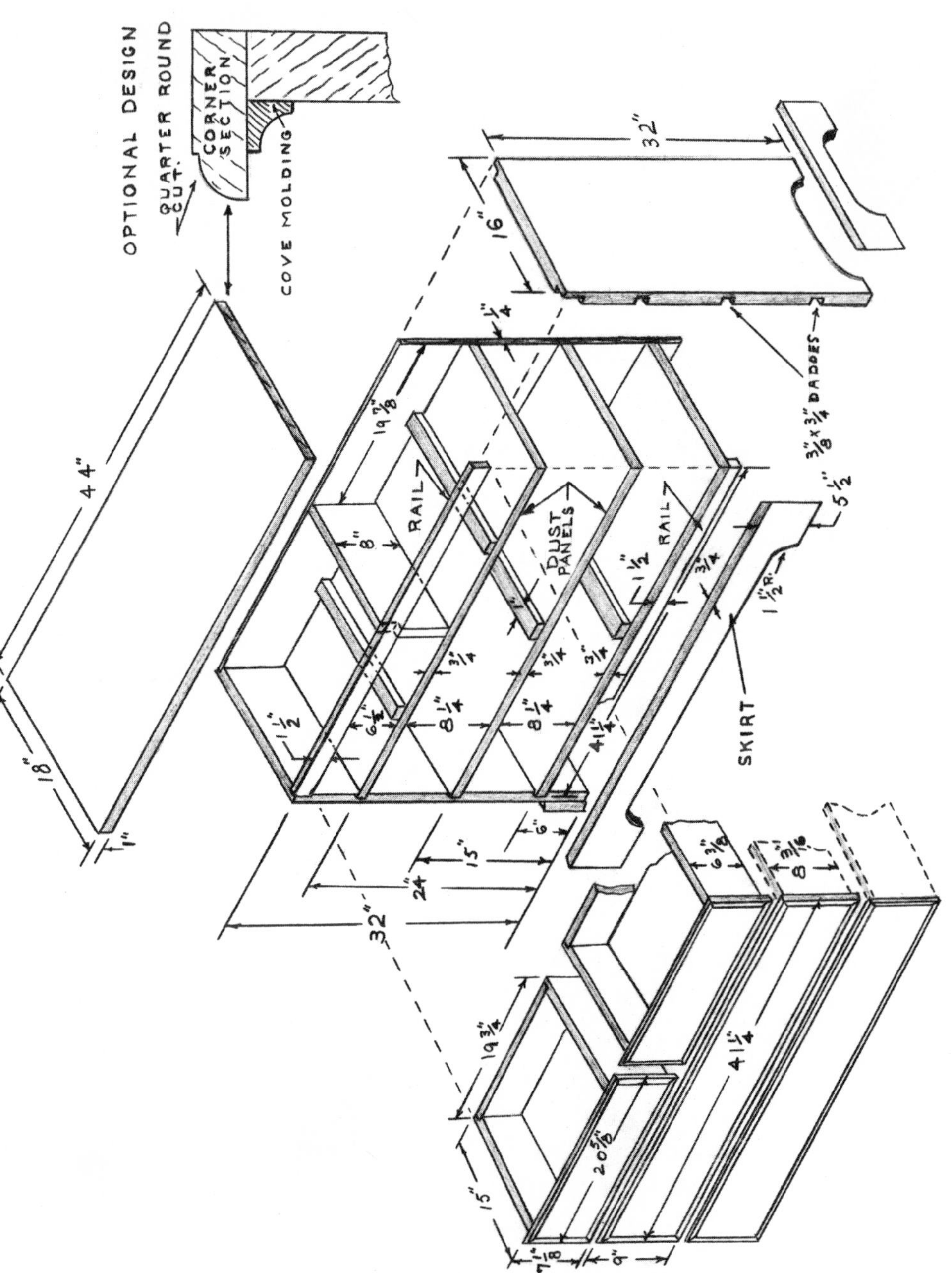

3. Cut square (90 degrees), to the size indicated, the two rails, three drawer divider dust panels, and the back.

   Plane, scrape, and sand all of these pieces for preparation of assembly.

   Insert glue into the dadoes, then insert all of the pieces into their matching joints using bar clamps for pressure. Check for squareness of assembly with a carpenter's square before the glue sets up.

   The top rail is attached to the stop rabbet joints with glue and #6 finishing nails.

   Install the ¼-inch back to the precut rabbets with ⅞-inch flathead brads.

   *Note:* If additional support is desired, nail #6 finishing nails through the dadoes and into the divider dust panels.

4. Lay out and cut a ¾-inch by 1½-inch stop rabbet into the front top edge of the top middle drawer divider.

   Attach the drawer divider to the exact center of the top drawer rail by gluing and clamping; reinforce with #6 finishing nails. Continue to fasten the top to the assembly by drilling counterbored screw holes then installing #10, 1½-inch flathead screws.

   *Note:* Make sure the screws are equally divided and aligned to each other. Make a quarter-round design along the three edges with an electric router. Make or purchase cove molding, then cut 45-degree mitred corners and fasten in place with #4 finishing nails.

5. Glue and clamp a ¾-inch by 1½-inch by 40½-inch rail to the underface of the bottom dust panel.

Lay out and cut the design for the skirt. The top edge of the skirt is attractive if cut at a 30-degree bevel.

Cut mitred corners for the three skirt pieces, then fasten them to the base of the assembly with glue and #6 finishing nails.

6. Lay out and cut the drawer guides. Nail them to the exact center of each solid divider dust panel. Refer to the section in the front of the book on constructing drawer guides and rides.
7. Construct each drawer unit by using the construction hints found in the front of the book. Make sure the front of each drawer has a ⅜-inch overlap along the inside edge. Make a slight bevel or chamfer cut on the four outside edges.
8. Set and fill all nail holes, then plug all counterbored screw holes. Thoroughly sand the bureau to a smooth surface, then stain or paint as desired. Rub with #420 emery paper or pumice and lemon oil. Preserve the finish with an occasional coat of lemon oil or paste wax.
9. Install the hardware of your choice. Brass or wood knobs, 1¼-inch diameter, are attractive.

## Project 76

# GUN CABINET

### Materials Required

*Wood of Your Choice*

**2 sides** ¾″ × 11″ × 64½″
**2 shelves** ¾″ × 10¾″ × 24″
**1 top** ¾″ × 11¾″ × 26½″
**1 door header** ¾″ × 1″ × 23½″
**1 front base skirt** ¾″ × 3″ × 23½″
**1 back** ¼″ × 24¾″ × 61″
**1 drawer unit (overall size)** 4½″ × 10⅞″ × 23¼″ with a ⅜″ lip around the front
**1 barrel holder** ¾″ × 2½″ × 23½″
**1 butt holder** ¾″ × 7½″ × 23½″
**2 door stiles** ¾″ × 2½″ × 53½″
**2 door rails** ¾″ × 2½″ × 24″
**1 piece of glass** ⅛″ × 19¾″ × 49¼″
**1 barrel cylinder lock for door**
**3 offset hinges** 2″ × ⅜″
**2 drawer pulls**

### Procedures

1. Lay out and cut the two sides to the dimensions given (¾ by 11 by 64½ inches).
2. Cut ¼-inch by ¾-inch dadoes on the inner faces of the side pieces to receive the two lower shelves.

   Cut ¼-inch by ⅜-inch rabbets on the back inside edges of the sides to receive the back.
3. Lay out and cut the two shelves to size. Be sure they are cut square.

   Lay out and cut the top to size. Shape the ends and front with a router and quarter-round bit.

# Gun Cabinet

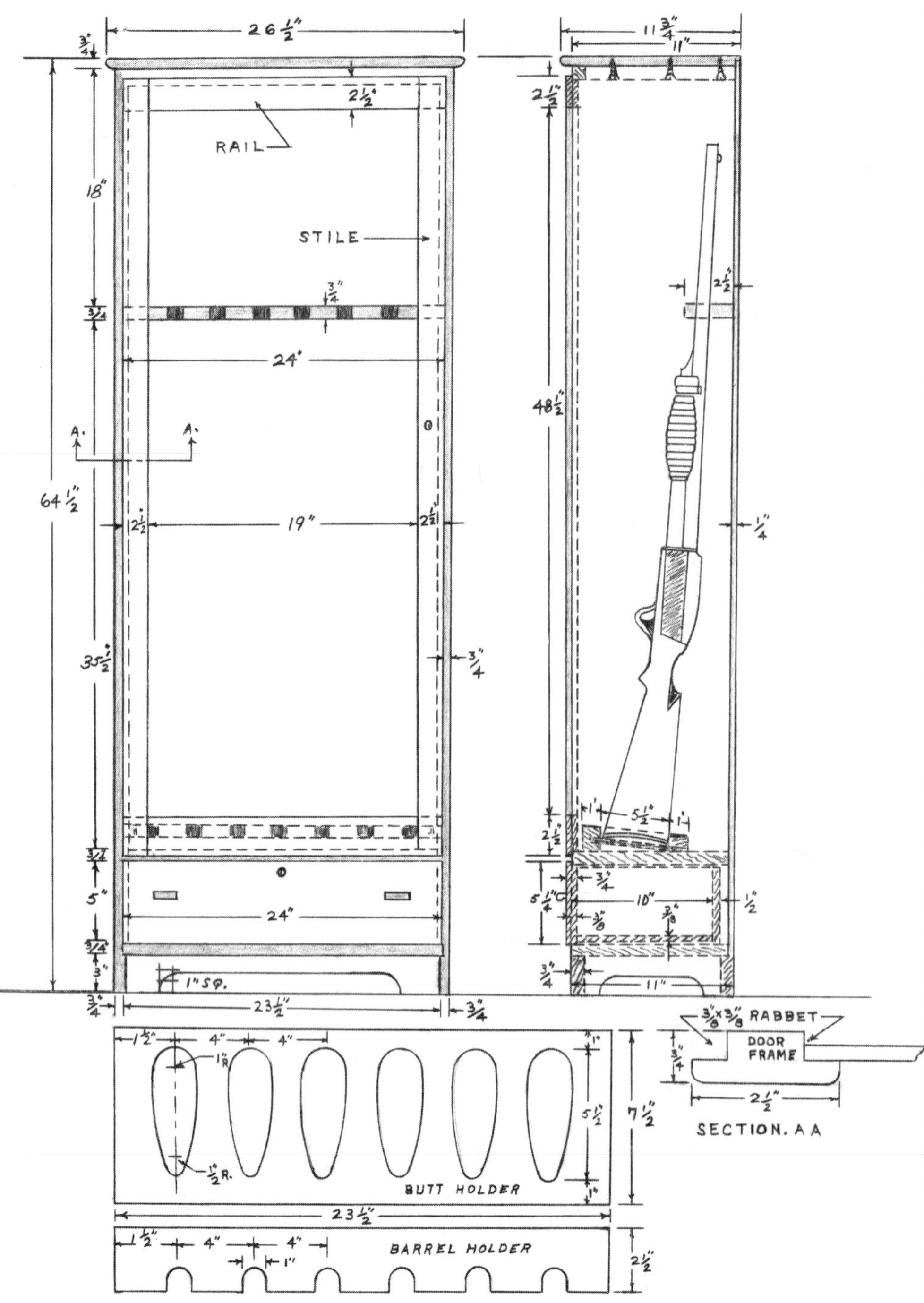

26 1/2"
3/4"
2 1/2"
RAIL
18"
STILE
3/4"
3/4
24"
A.
A.
64 1/2"
2 1/2"
19"
2 1/2"
35 1/2"
3/4"
3/4
5"
24"
3/4"
3"
1" SQ.
3/4"
23 1/2"
3/4
11 3/4"
11"
2 1/2"
2 1/2"
48 1/2"
1/4"
1"
5 1/2"
1"
2 1/2"
3/4"
5 1/4"
10"
1/2"
3/8"
3/8"
3/4"
11"
1 1/2"
4"
4"
1"
1" R.
5 1/2"
7 1/2"
1/2" R.
BUTT HOLDER
1"
23 1/2"
1 1/2"
4"
4"
1"
BARREL HOLDER
2 1/2"
3/8" x 3/8" RABBET
DOOR FRAME
3/4"
2 1/2"
SECTION. A A

Cut a (¼-inch by ⅜-inch) rabbet into the back edge of the top piece to receive the back.

4. Construct the barrel holder by boring 1-inch holes and sawing out the wood in front of the holes. File the openings smooth.

5. Smooth the pieces by scraping, planing, and sanding. Proceed to fasten the cabinet frame together by gluing and clamping with bar clamps.

   *Note:* Before the glue sets up, check the structure for squareness.

6. Lay out and construct the base skirt and the door header, then fasten them into position by gluing and clamping. Screws may be added for extra strength.

7. Construct the drawer unit to the overall size given in the drawing. Make sure the front has a ⅜-inch drawer to overlap the cabinet frame. Refer to drawer construction hints found in the front of the book. Round over or chamfer the outside four sides of the drawer.

8. Lay out and cut square the ¼-inch back, then attach it to the precut rabbets with ¾-inch brads.

9. Construct the butt holder by sawing out the holes on the design in the drawing with a sabre saw. Tilt the butt holder up by placing a 1-inch by 1-inch strip below the front edge. Cut the top of the strip to a 10-degree bevel.

10. Cut all parts to the door. Make end half-lap joints to fasten corners. Cut ⅜-inch by ⅜-inch rabbets on both inside and outside edges of each part to receive the glass and to fit into the cabinet door opening.

11. Fasten the door together by gluing and clamping the end half-lap joints. Round the inner and outer edge with a router or quarter-round bit.
12. Fit the door into position, then attach in place with the three offset hinges.
13. Bore the correct size holes, then install the two cylinder locks. Fasten the drawer pulls in place.
14. Sand the project smooth, then apply the finish of your choice, rubbing between coats with 320 emery paper. Preserve the finish with a coat of paste wax.
15. Install the glass in place with glass points, then cover over the rabbet cutout with ¼-inch by ⅜-inch molding.

## Project 77

# CHANDELIER

### Materials Required

*Wood of Your Choice*

**1 centerpiece** 3½″ × 3½″ × 7½″
**8 candleholders** 2″ × 2″ × 4″
**8 dowels** ⅜″ × 14″
**2 shaker knobs** 1¾″ dia.
**8 shaker knobs** 1¼″ dia.
**1 eye hook** ¾″

*Note:* An alternative to the shaker knobs is brass or porcelain knobs. These can be purchased at a local hardware store.

### Procedures

1. Lay out and cut all stock to the suggested size.

2. Lay out and cut the ½-inch chamfer designs on the centerpiece. Continue to lay out and cut the four chamfered designs on the eight candleholder blocks.

3. Position and bore the eight ⅜-inch-diameter holes into the centerpiece to a depth of ¾ inch. Be sure that all holes are aligned. Position and bore the eight ⅜-inch holes into the eight candleholders to a depth of ¾ inch.

4. Fasten the correct knobs to the centerpiece and candleholders by gluing and clamping. If the knobs have a tenon, bore the correct hole then fasten with glue.

# Chandelier

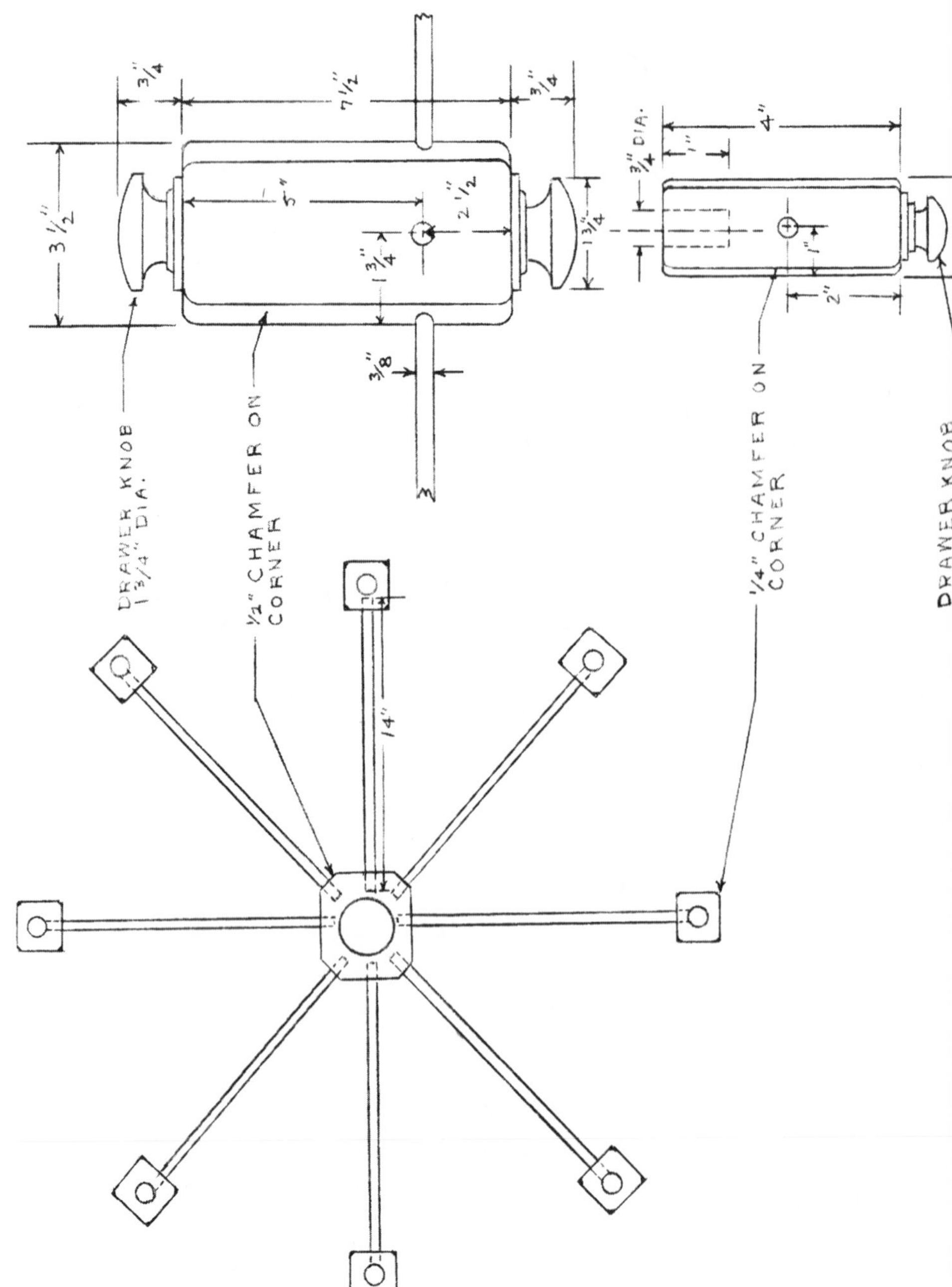
DRAWER KNOB
1 3/4" DIA.
1/2" CHAMFER ON
CORNER
1/4" CHAMFER ON
CORNER
DRAWER KNOB
3/4" DIA.
7 1/2"
3/4"
3/4"
3 1/2"
5"
2 1/2"
1 3/4"
1 3/4"
3/8"
4"
1"
1"
2"
14"

5. With glue fasten all the dowels into the holes in the centerpiece and candleholders. Be sure that the candleholders are aligned to one another.
6. Clean off all traces of glue and proceed to sand the project to a smooth surface. Since this project is made of hardwood, a clear finish is best suited. Apply several coats of finishing oil, rubbing lightly betwen each coat with fine steel wool or pumice and oil. To protect the finish use paste wax and buff.
7. Attach the ¾-inch eye hook to the top of the centerpiece.

## Project 78

# WOODWORKING BENCH

### Materials Required

*Wood of Your Choice*

**1 top** 2½″ × 18″ × 52″
**1 top (tool well section)** 1″ × 10″ × 52″
**1 tool well back support** 1″ × 2″ × 52″
**1 tool slot piece** 1″ × 2½″ × 52″
**3 blocks** 1″ × 2½″ × 3″
**4 legs** 2½″ × 2½″ × 29½″
**2 apron rails** 1″ × 4″ × 36″
**2 apron rails** 1″ × 4″ × 26″
**2 rails** 1″ × 4″ × 36″
**2 rails** 1″ × 4″ × 26″
**2 cleats** 1″ × 2″ × 29″
**2 cleats** 1″ × 2″ × 21″
**2 bench stops**
**1 vise**
**32 carriage bolts with washers and nuts** ¼″ × 4″

### Procedures

1. Lay out and cut all pieces to size, then begin by setting up bar clamps and edge-gluing enough stock to meet the required width for the bench top. It may also be necessary to laminate stock for the 2½-inch-thick legs.
2. Locate and drill the four ¼-inch-diameter holes through each apron and rail.

   Locate and drill the eight ¼-inch-diameter holes through each leg. Be sure to drill the holes so that they are angled slightly away from each other; this will prevent the bolts from hitting each other.

# Woodworking Bench

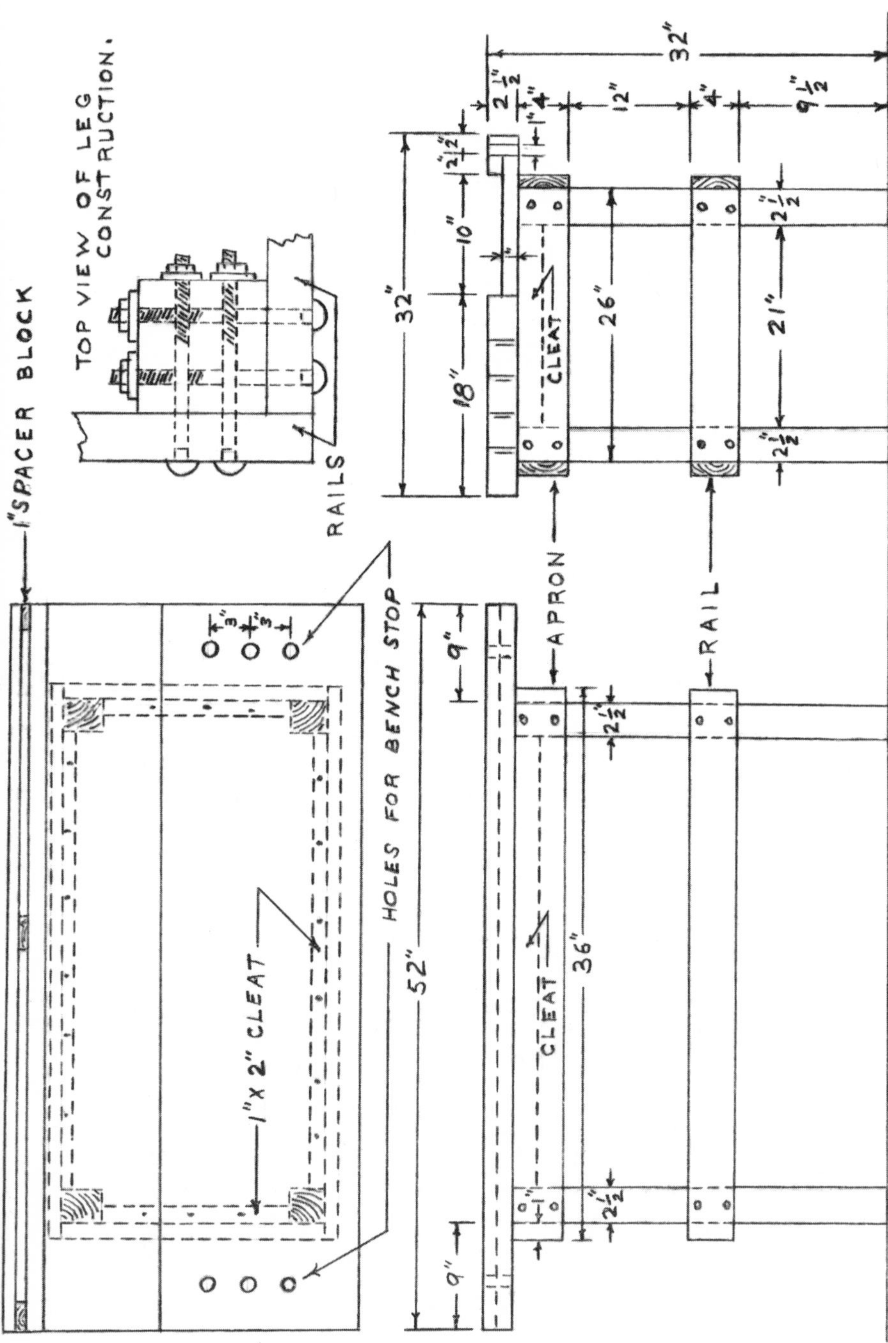

3. Assemble the base together by bolting the eight apron and rail pieces into the prebored holes in the legs. Fasten the 32 carriage bolts tightly with washers and nuts.

4. At approximately every 8 inches drill a countersunk pilot hole through each 1-inch by 2-inch cleat, then attach the cleats onto the apron so that they fit snugly between each pair of legs. Use #12, 1¾-inch flathead wood screws. The top edge of the cleats should be flush with the top edge of the apron pieces.

5. To fasten the top to the leg assembly from beneath the table, drill counterbored pilot holes up through the cleats and into the top at approximately 6-inch spacing. Use #12, 3-inch flathead wood screws to attach the top to the cleats.

6. Glue and screw the tool well back support to the rear of the tool well section. The pilot holes can be drilled from beneath the table.

   Fasten the three 1-inch by 2½-inch by 3-inch spacing blocks to the back edge of the tool well with glue and #6 nails.

   Complete the tool slot by nailing onto the spacing blocks the 1-inch by 2½-inch by 52-inch strip.

7. Locate and bore the six ¾-inch-diameter holes for the bench stops.

8. Fasten a 1-inch by 4-inch by 7½-inch vise block beneath the left or right end of the table with #10, 2-inch flathead wood screws. Mount the vise to the edge of the table and to vise block with flathead screws.

9. Finish the bench with several coats of oil.

WOODWORKER'S VISE
TOOL WELL
SLOT TOOL RACK
BENCH STOP

Project 79

# STEREO CABINET

## Materials Required

*Wood of Your Choice*

**1 top** ¾″ × 22″ × 54½″
**2 sides** ¾″ × 22″ × 36″
**1 shelf** ¾″ × 21¾″ × 32″
**1 bottom** ¾″ × 21¾″ × 54½″
**4 record dividers** ¾″ × 21¾″ × 23″
**1 section divider** ¾″ × 21¾″ × 35″
**1 slide-out shelf** ¾″ × 21½″ × 22½″
**4 shelf guides** ¾″ × ¾″ × 20″
**2 top cleats** 1″ × 1″ × 18″
**2 bottom cleats** 1″ × 1″ × 18″
**1 back** ¼″ × 35¼″ × 55″

## Procedures

1. If the stereo cabinet is to be built from standard hardwood stock, it will be necessary to use bar clamps to edge-glue enough stock to reach the required width for all pieces.

2. Lay out and cut all pieces to the size indicated in the drawing and the materials list.

3. Make ¼-inch by ¼-inch rabbet cuts on the rear inner edge of the top and two sides to receive the ¼-inch back. On the side pieces a blind rabbet should be cut so that it does not extend to the top edge.

# Stereo Cabinet

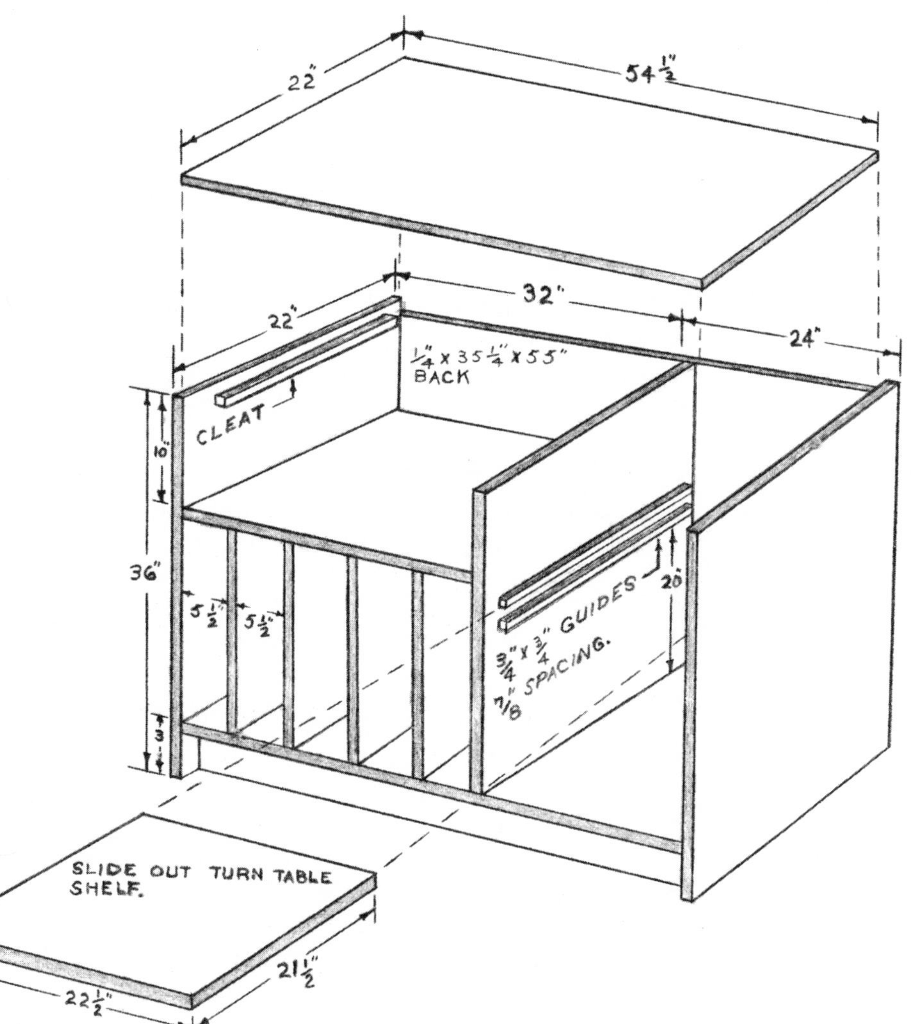

4. Fasten the top and bottom cleats to the inner face of each side piece. The top cleats should be placed 1 inch from the top edge of the sides, and the bottom cleats should be secured 2¼ inches above the floor line. The cleats are used to secure the top and bottom to the sides. Use glue and #8, 1½-inch screws to fasten.

5. To fasten the top and bottom to the sides, begin by positioning the top and bottom on the cleats with clamps, then drill countersunk pilot holes up through the cleats and into the underface of the two pieces. Fasten the top and bottom to the cleats with #10, 1½-inch flathead screws.

6. Make sure the assembly is square, then proceed to nail the ¼-inch back into the precut rabbets. Use 1-inch finishing brads.

7. Nail and glue the section divider into place, then fasten the shelf to the side and divider panel.

   Lay out and secure the record divider panels to the bottom and top shelf using glue and #6 finishing nails as fasteners. Nail and glue the kick board to the sides and bottom shelf.

8. For the turntable shelf to slide out correctly fasten two ¾-inch by ¾-inch by 18-inch guide strips to each side of the turntable section. The strips should be spaced approximately ⅞ inch to allow the ¾-inch-thick sliding shelf to move freely. Be sure the guide pieces are completely aligned. Insert the ¾ by 22½ by 22½-inch shelf between the guide strips.

   *Note:* Use wax or soap on the bottom guide strip to allow the shelf to move in and out freely.

9. Set and fill all holes, scrape off all traces of glue, then proceed to sand the unit to a smooth finished surface.

10. Select and apply the stain of your choice. When dry, apply four coats of clear lacquer, rubbing between coats with pumice and lemon oil. Preserve the finish with an occasional coat of lemon oil.

## Project 80

# MAIL BOX

### Materials Required

*Wood of Your Choice*

**1 back** ½″ × 9½″ × 14″
**2 sides** ½″ × 5¾″ × 6½″
**1 top** ½″ × 1″ × 14″
**1 lid** ½″ × 6″ × 14″
**1 bottom** ½″ × 4¾″ × 13″
**2 butt hinges** 1″
**nails** #4 finishing

### Procedures

1. Lay out and cut all parts to the suggested size and design.

2. Fasten the sides to the back, then fasten the bottom to the back and side pieces. Cut the top edge of the front piece to a 40-degree angle, then fasten the front to the sides and bottom. Use glue and #4 finishing nails.

3. Lay out and cut the top and bottom edge of the lid to a 40-degree angle.

4. Lay out and cut the four $^{1}/_{16}$-inch by 1-inch gains to receive the two butt hinges. Attach the two 1-inch butt hinges to the top piece and to the lid, then fasten the top piece to the top of the shelves with glue and #4 finishing nails.

5. Clean off all traces of glue. Set and fill all nail holes and sand the box to a smooth surface.

# Mail Box

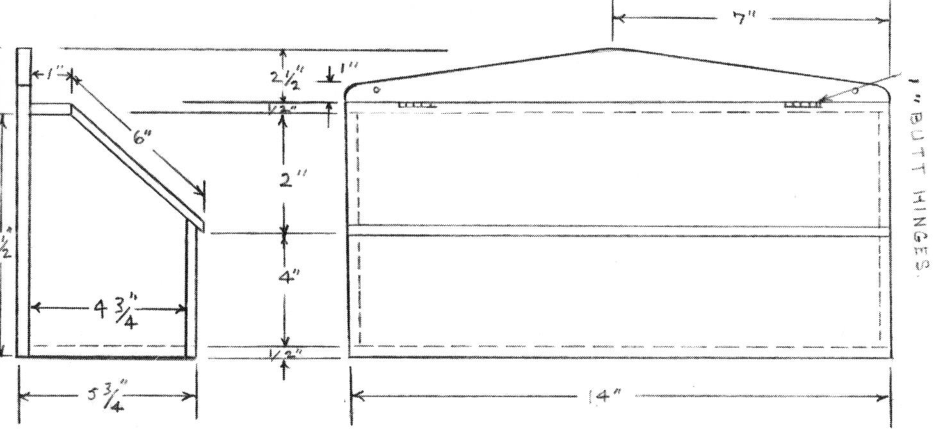


**6.** Select and apply the colored paint or stain of your choice. When dry, apply several coats of spar varnish, rubbing between coats with fine steel wool or pumice and oil. To protect the finish use paste wax and buff.

*Note:* Spar varnish is the best available finish for outdoor use, as it repels moisture and salt.

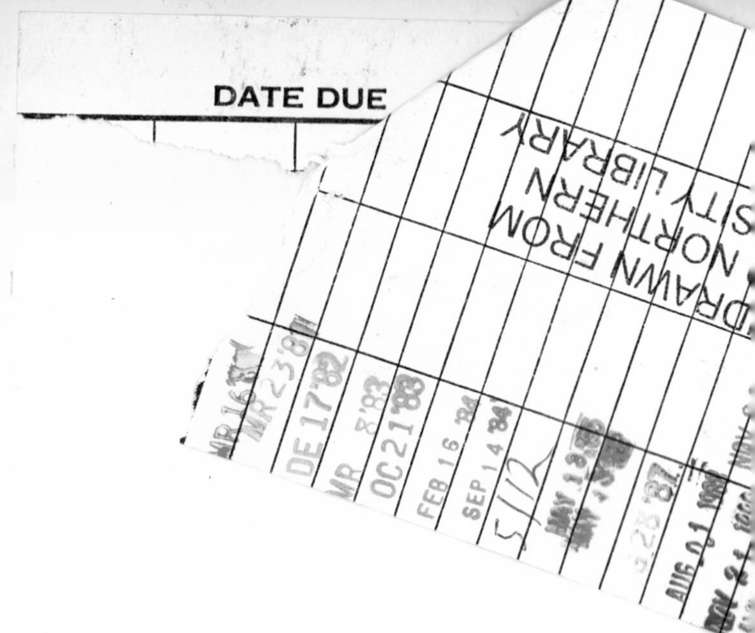

HETERICK MEMORIAL LIBRARY
684.1042 O93e onuu
Ouimet, Ronald P./80 woodcraft projects
3 5111 00077 6132